1 - 60

TECHNOLOGY AND PLACE

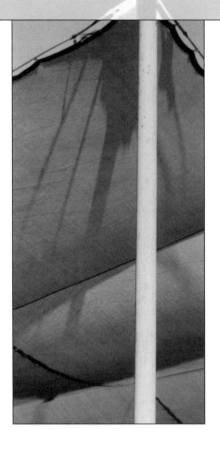

TECHNOLOGY

Foreword by
Kenneth Frampton

STEVEN A. MOORE

A N D   P L A C E

SUSTAINABLE

ARCHITECTURE

and the

BLUEPRINT

FARM

University of Texas Press Austin

The publication of this book was assisted by a University Cooperative Society Subvention Grant awarded by the University of Texas at Austin.

Requests for permission to reproduce material from this work should be sent to Permissions, University of Texas Press, P.O. Box 7819, Austin, TX 78713-7819.

♾ The paper used in this book meets the minimum requirements of ANSI/NISO Z39.48-1992 (R1997) (Permanence of Paper).

LIBRARY OF CONGRESS CATALOGING-IN-PUBLICATION DATA

Moore, Steven A., 1945–
    Technology and place : sustainable architecture and the Blueprint Farm / Steven A. Moore ; foreword by Kenneth Frampton.— 1st ed.
        p.   cm.
Includes bibliographical references and index.
    ISBN 0-292-75244-X (alk. paper) — ISBN 0-292-75245-8 (pbk. : alk. paper)
    1. Blueprint Farm (Laredo, Tex.)   2. Agricultural innovations—Texas—Laredo.
3. Technology—Social aspects—Texas—Laredo.   4. Farm buildings—Texas—Laredo.
5. Architecture, Modern—20th century—Texas—Laredo.
I. Title.
    S451.T4 M66   2001
    631.2'09764—dc21   00-010975

FOR MARJORIE

CONTENTS

FIGURES

BY KENNETH FRAMPTON

FOREWORD

It is one of the characteristics of architecture that as opposed to the other arts it is always contingent upon power. However idealized our image of architectural practice might be, architects are always made forcibly aware of the contingencies imposed by authority and of the resistance of the environment to its significant transformation. Thus architects are unable to escape the "thickness of things," no matter with what subtlety and cunning they rightly assert their artistic freedom which is simultaneously both real and contingent. Architecture is necessarily as tempered by the task-in-hand as it is often determined by the dictates of bureaucracy and by the equally prejudicial worldview of the client. Thus, as Alvaro Siza has put it, "architects are specialists in non-specialization," if for no other reason than that they have to take into consideration an infinite variety of seemingly extraneous forces. This may well explain why architects tend to become involved in interdisciplinary studies. Certainly this is true of Steven Moore, the author of this partly pragmatic, partly philosophical text, devoted to the theme of technology and place and dedicated in the first instance to a postmortem of the demise of Blueprint Farm, an eco-tech experiment established in Laredo, Texas, in the second half of the 1980's, as an integrated study of the mutual sustainability of architecture and agriculture.

This ecological exercise, subject to unrealistic expectations, never had a chance of proving itself since it was prematurely aborted even before it became fully operational. Moore reveals how Blueprint Farm was the ambiguous consequence of ambitious democratic politics, miscommunication, bad faith, incompetence, and outright opportunism, culminating in its draconian closure largely made in the interests of subsidized, large-scale agriculture. I am alluding, of course, to the agribusiness as this continues to undermine medium-scale farming throughout the North American continent, while polluting non-renewable resources in order to maximize agricultural production. As an experiment in alternative agriculture, Blueprint Farm was handicapped from the start. It was unfavorably sited on poor agricultural land and inadequately funded at either the federal or the state level, not to mention the consortium of ill-matched non-profit foundations and agencies that were patched together without having any clear mandate or chain of command. As the author points out, there was a further irony in the fact that while the Israeli irrigation technique applied on the farm was first developed at Texas A&M, it had been perfected as a method by the Israeli state, as an integral part of its state-supported, socialist *kibbutzim* system.

Despite its premature abandonment, Blueprint Farm would prove to be extraordinarily productive as far as the theoretical side of this essay is concerned, serving as a basis for an altogether wider debate as to the cultural and political predicament facing our species as we enter the first decade of the twenty-first century; namely, how and by what means should the instrumentality of technoscience come to be mediated in the name of a broader concept of human culture and welfare? This is surely the ultimate stake in a critical moment in which agriculture and architecture, among other social practices such as medicine and mass transportation, are but subsets that are contingent for their long-term effectiveness on the political vision of the society as a whole.

Given Moore's declared aim that the ultimate intent of *Technology and Place* is to re-cast my own thesis as to the resistant potential of Critical Regionalism, it is surely somewhat invidious for me to assume the task of prefacing this essay.

Influenced by Fredric Jameson's critique as this appears in his book *The Seeds of Time* (1994) to the effect that Critical Regionalism is an unduly aestheticized hypothesis that is neither *modern* nor *postmodern*, Moore advances the concept of the *nonmodern* as a general modus with which to evade the maximizing thrust of technoscience that has now expanded to such an extent as to appear at times to escape control. This is especially evident, say, where both production and jobs have been exported beyond the reach of democratic process and regulation, particularly to third world countries, where labor is cheap and where control over the conditions of production and even the quality of product are all too easily circumvented. While this kind of transnational subversion of trade unions and labor laws does not as yet obtain to the same degree in the production of built environment, we have little reason for complacency given the way in which the "received" motopian network dominates the society to such a degree that the mutually symbiotic relationships obtaining both between people and things and between people and people are equally determined, not to say disaggregated, by the all-pervasive presence of the automobile.

Moore sees electrification as an earlier version of the same kind of hegemonic technological system, first brought into being as a universally available grid by Edison and subsequently developed in such a way as to eclipse gas as an alternative source of energy. Moore informs us as to the efficiency of gas-absorbing refrigerators which, much like gas cookers, consume less energy than those driven by electricity. That which obtains in the case of gas, natural or otherwise, also surely applies to many other alternative forms of eco-technology, some of which were attempted in the case of Blueprint Farm, such as wind-turbines, drip irrigation, rainwater collection, and the building of cheap, well-insulated recyclable buildings out of straw bales. Clearly the conservation of rainwater and the concomitant reduction in storm water erosion will become crucial factors in offsetting the future nemesis of suburbanization; namely, the pending worldwide scarcity of water. One should also perhaps note equally valid "eco-tech" approaches such as the recycling of "grey" water and warm air, tapping landfills for natural gas, and the related possibility of powering auto-mobiles with hydrogen, methane, or ethanol, particularly where the last two may be obtained through the fermentation of biomass.

The key to all this is the synergetic hybridization of available techniques, and here we have to recognize that special interests are always involved in the maximization of any given technology, as in, say, the over-use, not to say abuse, of antibiotics in allopathic medicine. Maximization clearly militates against the judicious application of those techniques that are most effective in socioecological terms, within the parameters of most problem situations, for it tends to preclude the possibility that a given set of needs and desires may be met by a more sustainable interaction between different techniques, such as the combination of light-rail transit with automotive distribution or the mediation of air-conditioning through natural ventilation and the use of double-glazed solar walls.

Paradoxically influenced by Heideggerian ontology and Marxist critical theory, Moore partially grounds his *nonmodern* discourse in the work of two late-modern thinkers who in distinctly different ways chose to distance themselves from the traditional Marxist theory of alienation, which tends to preclude life-enhancing, ameliorative strategies under the prevailing conditions of monopoly capitalism. The two figures are Henri Lefebvre in his books *The Critique of Everyday Life* and *The Production of Space,* dating from the 1960's, and Andrew Feenberg in his *Critical Theory of Technology,* published in 1991. Moore follows Lefebvre in insisting that space is not just a palpable, volumetric condition established within the built environment but rather a set of relations between things and their users. Under this rubric space/place comes to be embodied through sociotechnical practice rather than through the presence or absence of an aestheticized object. So much we may say for the late modern obsession with compensatory, admass consumption rather than the drive toward the life-enhancing potential of "lived-in" space.

By a similar token, Moore sees Feenberg as articulating a viable alternative to the social constructivist preoccupation with effecting standardized production through universal contractual agreements, since this purview tends to ignore the unequal power relations built into the adoption of state- or market-driven technological systems. Feenberg is in favor of a participatory work-place democracy which would moderate the technoscientific tendency to determine production solely in terms of efficiency. As far as Feenberg is concerned, market-driven societies suffer from a perennial failure of imagination in not being able to entertain alternative technological practices that are both efficient and life-enhancing. In short, Feenberg, following Herbert Marcuse, envisages a holistic ontology in which the future culture and welfare of the species would depend less upon an exploitative and technologically maximizing relationship with nature. For Feenberg, contemporary technoscience as a social practice is compromised by a false rationality, which is virtually anti-ecological by definition.

Two other figures are incorporated into Moore's eclectic construction of the *nonmodern*. The first is the philosopher Gianni Vattimo, whose idea of "weak thought" is categorically opposed to the technological hegemony of the Enlightenment, while not being stylistically antimodern or addicted to the compensations proffered by Guy Debord's "society of spectacle." The second seminal figure, from whom Moore actually derives his notion of the "nonmodern," is the sociologist Bruno Latour, whose texts, *Science in Action* and *We Have Never Been Modern,* cause Moore to aspire to Latour's activation of the social as a potentially liberative, non-transcendental view of technological development. Latour is surely at his most critical in questioning the technoscientific construction of reality, inasmuch as the web of interests that support a given technique cannot be effectively challenged from the outside once the instrumental facts have been legitimized, as it were, by a "black-box" scientific peer review.

Taking into consideration the various "disalienation" theses advanced by Lefebvre, Feenberg, Vattimo, and Latour, Moore aligns himself with the synthetic "nonmodern" work of the American landscape architect John Tillman Lyle and his concept of a "regenerative architecture," as this is set forth in his book *Regenerative Design for Sustainable Development* of 1994. It is hardly surprising, given Lyle's professional formation, that the larger part of his study should address itself to the cultivation of landscape in terms of both energy flow and hydraulic conservation to which all land-forms must react even in the driest of deserts. Many pages are thus devoted, in pragmatic terms, to recommended schemes for the conservation, absorption, diversion, and filtration of storm water, although this by no means exhausts the full scope of Lyle's comprehensive ecological know-how, which far from being a romantic critique of technoscience makes a very measured assessment of the pros and cons of various alternative strategies for the generation of energy and the cultivation of food. When it comes to architecture (and here one notes that Lyle much prefers the socio-biotechnical term *habitat*), it gradually becomes clear that we have been over much of this ground before, above all in the writings of Ian McHarg and also perhaps in the work of Frank Lloyd Wright, whose Kropotkinian vision of Broadacre City was surely moving in a regenerative but nonetheless highly cultured direction as long ago as 1934.

This is where Moore ostensibly leaves us with the topographic proto-ecological implications of Critical Regionalism, sharply divested of its aesthetic, psycho-sensual, and anarcho-social conceits. And yet we are still brought back to confront the double aporia of our admass time: first, how do people get to want what they want (the paradox of desire, its arousal and gratification) and, second, how does the *vox populi* get what it finally opts for in terms of the current processes of our liberal democracy? This is the essential dilemma posed by rationality and power in the late-modern world, be it post-socialist or otherwise, and it is to this we have to turn as architects and non-architects alike, for in the last analysis everything rests with the client in the largest possible sense of the term. Here Moore returns us to the theme of education, not in the professional sense of the term but rather at the level of the society as a whole and not solely with respect to higher education but rather first and foremost in terms of the average high school curriculum. For until we get a higher level of awareness about the full scope of the environmental crisis induced by the relentless drives of our technoscientific civilization, very little progress will be made towards a consensual acceptance of a regenerative architecture in which building, dwelling, and cultivating are once again one and the same.

PREFACE

For the past thirty years or so, architectural discourse has been dominated by aesthetic concerns. In what follows I do not argue that we should abandon our love of the aesthetic object, the beautiful, the poetic, or even the visual. I argue only that we should balance our fascination with the world of appearances by remembering that architecture is an ecological, technological, and political *practice*. This is a work, then, that will attempt to influence theory by empirically examining the conditions of contemporary practice.

My interest in the tension between theory and practice is, of course, personal as well as scholarly. Having practiced architecture for twenty years before entering academic life, I found that much of the academic literature I encountered, while of intellectual interest, simply ignored the conditions of architectural production. Having practiced for so long, I had become only too familiar with the everyday political and material demands of building in the public arena. It didn't take me long to figure out that my formal study of architecture would necessarily bridge the ever-widening gulf between those who interpret construction and those who construct.

I also came to the academy with a growing interest in the emerging concept of "sustainability." Ten years ago I was at a loss to explain why the concept of ecology had been so marginalized within the discipline of architecture and within society as a whole. Being thus perplexed by two related topics—practice and sustainability—I thought it only natural to select a case for study that involved an ecologically based public controversy. It seemed to me then, as now, that such a study might enable us to better understand how and why our society chooses to build and live as it does.

This study was, then, first conceived and written as a doctoral dissertation. That format requires at least two things. First, it requires a certain density of content that demonstrates a broad understanding of the literature in the field(s) of investigation. In this iteration of the project I have purged a lengthy section on methodology and have attempted generally to translate the remainder into language that will be accessible to most readers. Some, however, will complain that sections of the remaining text are still impenetrably dense. To those readers I apologize for my inability to communicate ideas more simply.

The second requirement of academic study is to rigorously protect the identity of those private individuals who contributed so generously to my understanding of the case. The names of public figures such as Jim Hightower, Pliny Fisk III, and Gail Vittori are not concealed because they are a matter of public record. The names of many other individuals used in this text are, however, pseudonyms because I am guarding the privacy of sources. Such citations are drawn from twelve in-depth interviews gathered between May of 1995 and February of 1996, and several follow-up interviews that took place in the months and years following. Should any reader wish to verify the accuracy of the citations, I will make notes and tapes of interviews available for examination at the University of Texas at Austin. The names of respondents will, however, remain anonymous.

I am reluctant to describe this investigation as an ethnography because the period of my field study was, I imagine, something less than an anthropologist would require. I will say, however, that my study has been ethnographic, in that I have attempted to tell the story of Blueprint Farm in a way that will be familiar to those who built it and those who received it. It has, however, been impossible to make happy every party who has something at stake in this story. The very nature of this, or any, public event is that there are multiple, conflicting versions of reality at work. My hope is that this narrative, if not always flattering, will help to realize the noble intentions of Blueprint Farm's builders—if not at Laredo, then elsewhere.

ACKNOWLEDGMENTS

In academic publications it is customary to acknowledge the support of one's family at the end of this section. In this case, however, I must acknowledge that it was Marjorie Moore who enabled the beginning, the middle, and end of this work. At the beginning of this project her support was the catalyst for my departure from the full-time practice of architecture. That phased exodus enabled me to accept the Loeb Fellowship for Advanced Environmental Research at Harvard's Graduate School of Design, where my serious study of regionalism was begun in 1990–1991. At the end of this project it was also Marjorie Moore who made me aware of the MacDowell Fellowship. It was at the MacDowell Colony at Peterborough, New Hampshire, where I completed the manuscript in May of 1999. In between that beginning and end, intermittent production was enabled by her daily support.

Along the way I received the valuable support of mentors, colleagues, students, and other institutions. It is to Kenneth Frampton that I owe my greatest thanks for his mentorship. Not only has his scholarship stimulated my own, but he has been remarkably generous in his reception of my attempt to renovate his critical regionalism hypothesis. At Texas A&M University I have Frances Downing, Malcolm Quantrill, Paul Thompson, and Jonathan Smith to thank for their guidance. These scholars helped to shape the doctoral dissertation that preceded this book. Kathryn Henderson, Vince Canizaro, and Edward Burian, also of Texas A&M, provided criticism of early drafts that was invaluable.

At the University of Texas many colleagues have contributed to the development of the project in its current form. Bob Mugerauer and Michael Benedikt in particular have suggested specific revisions that have helped to clarify my nonmodern thesis. Stephen Ross was forever available to field and criticize evolving ideas, and Christopher Long helped to distill the proposal that led to this publication. Patricia Wilson provided a much-needed critique on the sections related to border studies. The graduate students in my philosophy and history of technology seminars, particularly my graduate assistant, Iain Kerr, were more influential than they realize in helping me test various propositions.

Scholars at other institutions were equally helpful. Stephen Fox of Rice University demonstrated remarkable patience in his criticism of the chapter "The Local History of Space." It was, however, Barbara Allen, director of the Science and Technology Studies Program of Virginia Polytechnic Institute and the State University of Northern Virginia Graduate Center, who helped most to sharpen and clarify the central arguments of the project. Under her editorship, Chapter 5 of the current book first appeared in the *Journal of Architectural Education* (JAE), published by the MIT Press, to which I am grateful for permission to print this revised text. The *JAE* also provided an opportunity for me to coedit, along with Kenneth Frampton, a complete issue of the journal devoted to the theme of this book, *Technology and Place*. I am grateful to those authors who contributed to that issue, and to my evolving understanding of the topic's breadth. That issue also includes my article "Technology, Place and the Nonmodern Thesis," which is derived from this text.

Financial support for research was made available by the American Institute of Architects and the American Architectural Foundation in the form of an AIA/AAF Advanced Research Scholarship, as well as by the Mike Hogg Endowment for Urban Studies, the University of Texas, and Texas A&M University. I am grateful for their support.

In the end, however, there would be no story to tell without the generous support of those who actually constructed and received Blueprint Farm. To the twelve respondents who are identified with pseudonyms in this text I owe much gratitude for their remarkably candid descriptions of the farm's production and their assessment of that process. To Pliny Fisk III and Gail Vittori, codirectors of the Center for Maximum Potential Building Systems, I owe even more gratitude. In an era when architectural criticism is too often debased by self-promotion and the interests of the industry, Fisk and Vittori have been both generous in their support of this project and receptive to the critical investigation of their work. They are a model for others to follow.

TECHNOLOGY AND PLACE

CHAPTER ONE

A QUESTION OF CATEGORIES

The old word *bauen,* which says that man *is* insofar as he *dwells,* this word bauen however also means at the same time to cherish and protect, to preserve and care for, specifically to till the soil, to cultivate the vine. Such building only takes care— it tends the growth that ripens into fruit of its own accord. Building in the sense of nurturing is not making anything. Ship-building and temple-building, on the other hand, do in a certain way make their own works. Here, building, in contrast with cultivating, Latin *colere, cultura,* and building as in raising up edifices, *aedificare*— are comprised with genuine building, that is, dwelling.

Martin Heidegger, "Building Dwelling Thinking," pp. 146–147

The German philosopher Heidegger … based his allegiance to the principles (if not the practices) of Nazism on his rejection of the universalizing machine rationality as an appropriate mythology for modern life. He proposed, instead, a counter-myth of rootedness in place and environmentally-bound traditions as the only secure foundation for political and social action in a manifestly troubled world.

David Harvey, *The Condition of Postmodernity,* p. 35

Behind the assumptions implicit in the postmodern view of nature that emerges from Heidegger, and the assumptions implicit in the modern view of nature held by Harvey, is a conflict very much present in the construction and ironic conclusion of Blueprint Demonstration Farm.[1] This project was jointly developed by the Texas Department of Agriculture, the Center for Maximum Potential Building Systems of Austin, Laredo Junior College, and the Texas-Israel Exchange as an experiment in sustainable agricultural and architectural technology for semiarid ecosystems. Although the project has achieved almost cult status among those who support sustainable technology on a global scale, the project failed to develop a community of local supporters. As it neared completion in 1990, the state suddenly withdrew operating support, the Israelis retreated, and Laredo Junior College locked the gates. The project then languished as an archaeological ruin until 1995, when the college leased the site to a nonprofit organization unrelated to agricultural production. In 1998 the site was renovated for educational purposes. Thus the farm project represents a tragic breakdown between the intentions of those who *produced* it and those who *received* it. To understand the irony of this small story, and why it has big implications, it was not enough to study the farm buildings formally, or at a distance. Rather, I had to talk to people, and read their files. I had to engage myself in the place itself before the theories of others would be of much help in interpreting what happened in Laredo between 1987 and 1991.

Consider this small excerpt from the story:

In the fall of 1996, the Attorney for Public Affairs of the Texas Department of Agriculture reluctantly retrieved from the department's archive a black cardboard box. The opening of that black box revealed a file which documented that on the afternoon of June 6, 1989, "all the project players sat together in a meeting to clarify roles. The main purpose was to clarify that Pliny [Fisk III] had no recognized role in anything to do with agriculture for the project." In order to restore its waning authority over the construction of Blueprint Farm, where so much political goodwill was at stake, the Texas Department of Agriculture (TDA) found it necessary, publicly and loudly, to define the role of Fisk as the designer of Blueprint Farm, not its farmer.

The previous few months at the Laredo site had been ones of increasing institutional confusion and interpersonal rancor. After two years of planning and construction, things were administratively out of control. TDA convened the June meeting with its collaborators to assess the deteriorating situation and, once again, get the project under control. Tempers had flared, construction was behind schedule, and the budget was again strained. The crisis meeting was, in effect, called to issue a public cease and desist order—from that moment on it would be understood that architects do not farm and farmers do not build!

The categorical separation of *farming* and *building* at Blueprint Farm had actually been established, for accounting reasons, through an agreement among the Texas Department of Agriculture, the Meadows Foundation, and the Center for Maximum Potential Building Systems (CMPBS) at the incep-

Figure 1.1. Blueprint Farm as it appeared in 1995. Author's photograph.

tion of the project. The fateful June meeting itself was only the public manifestation of a political decision made earlier at a higher level. By invoking the
categories of accounting, TDA hoped to reestablish political order over the
"farm project" (which was controlled by Benni Gamliel, the Israeli farm manager) and the "building project" (which was controlled by codirectors Pliny
Fisk III and Gail Vittori of the Center for Maximum Potential Building Systems).
The administrators of the project understood that with control over cash flow
comes authority. They assumed that what couldn't be controlled by policy
management could be controlled by money management.

A year and a half later, in December 1990 (when the building project was
still not complete), the designer was again caught "farming." A new project
administrator, Dean Donald Hegwood of Texas A&I University, denied Fisk's
request to continue with a composting program that recycled residential and
campus garbage as crop fertilizer to be used on the farm. Hegwood reasoned
that such "organic" activity was a part of crop production and not a "construction" activity. Hegwood may not have known that Pliny Fisk II, the designer's
father, holds several U.S. patents on composting technology. It is also not likely
that he cared. What Hegwood did know and care about was that his administrative categories were being violated and that it was his job to keep accounts
straight and his Israeli colleagues happy.

As Hegwood once again attempted to restore administrative order on
the construction site, Dr. Héctor Jiménez watched through the window of his
second-floor office across campus with palpable disappointment. At the beginning of the project, he had great expectations for the demonstration farm.
His understanding was that the Israeli agricultural "scientists," and their American counterparts, were invited to Laredo to build something that would overcome the traditional patterns of rural poverty and patrimonial land-use practices that are deeply rooted in the culture of la Frontera Chica—the border
region shared by the United States and Mexico across the Rio Grande and
centered in Laredo. Instead of making the desert bloom, however, these mere
"practitioners," as Dr. Jiménez derisively referred to them, seemed to do noth-

ing but bicker about chemicals and turf—these were both literal and figurative turf battles.[2] It was discouraging. It seemed to Dr. Jiménez that the potential of science and technology to solve la Frontera Chica's deep economic and social problems was being sabotaged by hair-splitting philosophical squabbles. Alvaro Lacayo, a local farm and labor activist, described the emerging ideological space between the "farm site" and the "building site" as the "DMZ"—thus charging his description of the place with a Vietnam War–era reference to a demilitarized zone.[3]

Lacayo's military metaphor was ironically suited to this landscape of technological conflict. On one side were those like Benni Gamliel, the Israeli farm manager who wished to maximize production by employing the rational methods of modernization. On the other side were those like the Texan designer, Pliny Fisk III, who were more interested in how organic production might transform the lives of local farmworkers. For these *ecologists,* however, the making of the farm must have been understood as an end in itself because nothing seemed to ever get finished.[4] From Dr. Jiménez's point of view, the controversy between those who supported organic technology and those who supported advanced technology only prevented real progress from being made.

When the farm was eventually closed in 1991, Jiménez understood the tragedy as an agreement, not about what should be done to change social conditions in la Frontera Chica, as he had initially hoped, but as an agreement between ideological combatants to suppress the problem lived by his community. On Jiménez's account, Blueprint Farm is a story of *technology, place, and irony.* It is a story about technology because an unlikely group of collaborators imagined that technology itself might alter the trajectory of history. It is a story about place because the conditions into which technology was thrown proved to be unique and resistant to grand plans drawn up at a distance. And finally, it is an ironic story because the sustainable technologies deployed to change the lives of displaced farmworkers produced conditions that were the opposite of those intended. The farm was not *sustained.*

This small scene, and the story to follow, suggest that the ironic case of Blueprint Farm is relevant to a broader inquiry regarding the future of architectural theory and practice. The inability of some collaborators at Blueprint Farm to recognize a historical relationship between the practices of architecture and agriculture is emblematic of conventional modes of interpreting the world. Countless other events that took place at Laredo between 1987 and 1991 are equally salient. Reconstructing these conflicts will provide the empirical evidence needed to construct the theoretical arguments put forward below.

In sum, I argue that the events that took place at Blueprint Farm exemplify the larger structures embedded in the modern/postmodern opposition implied in the passages by Heidegger and Harvey cited at the opening of this chapter. That we can distinguish between modern and postmodern interpretations of reality is a philosophical condition made concrete by the conflicting interpretations of this case. How, then, shall we understand this Texas tragedy: through postmodern or modern lenses?

Heidegger's passage reestablishes a linguistic memory between the act of "edification" and the "building of soil"—between architecture and agriculture. I am not suggesting that all architecture is somehow related to farming, or the reverse, that all farming is architectural in the conventional usage of those terms. I am arguing, however, that architecture and agriculture share an archaic history that, as Heidegger argues, "preserves and cares for" the earth. The very idea of an ecologically based architecture begins with the critique of modern technology that is exemplified, if not originated, by Heidegger's postmodern text. Harvey's passage, however, reminds us that the politics of ecologism are never simple. He argues that the "rejection of . . . universalizing machine rationality as an appropriate mythology for modern life" too easily leads to a nostalgia for Being that only recuperates those social hierarchies that have propped up social injustice for too long. Anna Bramwell goes even further. Although she overstates her case, she presents credible evidence that documents an historical association between ecologism and radical fascism.[5] To argue, then, that the real events of the farm are merely the *shadow* of a purely theoretical discourse reasons that to privilege Being—the spatialization of human relations—over becoming—the decontextualization of human relations by time—is to fall into the philosophical trap which he imagines to have been set by Heidegger. Harvey finally embraces the radical "unreality of place" as the false consciousness of those who oppose the "unity between peoples." The unreality of Yugoslavia as a monolithic place-bound community may serve as a tragic example of Harvey's worst fears. Although bounded places may be familiar and comforting to those who identify with the "law of the land," the (re)description of boundaries may operate as a device to exclude and cleanse Others from our midst.[20] The cultural politics of region-making are rarely benign because the categories we impose upon nature have been so often employed ideologically to categorize humans as Others.

Among modern architects, Le Corbusier was perhaps the most vehement critic of regionalist place-based sentiments. In 1925, Le Corbusier complained that the "picturesque regionalism" of Camillo Sitte and "this touching renaissance of the home ... [were] destined grotesquely to divert architecture from its proper path."[21] Le Corbusier's ironic reference to the "home" is a classic example of the reification of traditional moral codes by the ideologues of modernism. His reference to the "proper path" of architecture is likewise a reproduction of the modernist teleology of history. Le Corbusier's internationalist program embodied precisely those attitudes that Agnew has identified as having devalued the concept of place in modernist thought.

Postmodern discourse, however, would invert Le Corbusier's program. Thinkers as dissimilar as the philosophers Richard Rorty and Gianni Vattimo argue that we architects must give up our first principles, our gods, our master plans, and return to the understandable intimacy of local conditions. They argue that the totalizing projects of modernism are deeply, and inherently, flawed. To imagine we have the rational capacity to plan human societies in the abstract conditions of Cartesian space is not a symptom of modern injustice and chaos; *it is the cause.* In this postmodern view, the socialist master plan of Yugoslavia has resulted, not in the "unity between peoples" (as Harvey and Marshall Tito had hoped), but in the tribalization of place as ethnic territory. The tragedy of the former Yugoslavia is only the return of the repressed.

The modern privileging of technology over place has devalued regionalism, only to see local values reemerge as a powerful force in conservative postmodern architecture. In the view of Alexander Tzonis and Liane Lefaivre, architecture has served as the "memory machine" of political history—a device that "triggers" the familiar events of collective memory.[22] In this context, regional architecture has been a conservative (and sometimes positive) force, resistant to the a priori plans of internationalism. To the degree that postmodern formulations of place rely solely upon cultural familiarity, however, they serve to reproduce the conservative mission of place—to prop up the authority of self-selected aristocrats with the doctrines of environmental determinism.

My point here is that conservative postmoderns have appropriated, and merely inverted, the logic by which moderns have devalued the concept of place. First, the conflation of place and community has not been untangled by postmodern thought—postmoderns have simply reclaimed the moral codes that have been so objectionable to moderns. Second, although progressive postmoderns (such as Gianni Vattimo) have abandoned the very idea of history, conservatives (the architect Leon Krier, for example) have simply adopted *neo*traditional ideology in lieu of *anti*traditional ideology.[23] No regions, all regions, what's the difference? To remain within the force field spun by modernity, and to invert the poles, is only to be *anti*modern, not truly *post*modern.

CRITICAL
REGIONALISM
To argue, as I have done above, that place means something quite different from community is to make a helpful contrast, but not to adequately define the concept. Nor have I attempted to define what I mean by the term technology. Both of these concepts will be more rigorously defined in Chapter 3. For the moment it will suffice to argue that *technology* is a concept far more complex than *tool,* and that *place* is a concept far more complex than *locale.* Both places and tech-

nologies operate at a variety of scales, from the micro to the macro, and the concrete to the abstract. My intention in this introduction is only to set the stage for a theoretical position through which these concepts are better understood. The doctrine of critical regionalism provides a platform for such improved understanding.

This doctrine proposes a major renovation to modernist ideology that avoids such a simple binary opposition between places and technologies.[24] The salience of Blueprint Farm to the theoretical problem pondered above is that this project lends itself to investigation as a case of critical regionalism in architecture. It does so because it defies the modern/postmodern opposition of categories illustrated in Figure 1.2. As this inquiry will make clear, the producers of Blueprint Farm intended to both utilize experimental technologies *and* respond positively to the particular qualities of place. It is the social complexity of this attempt that makes Blueprint Farm such a compelling story.

Kenneth Frampton, the principal progenitor of critical regionalism, argues that local places, rather than being "coercive, limiting, or idiotic," as Marx understood them, might be constructed and lived as both positive and progressive forces. In this sense, the local limits of places are considered constructive when they are resistant to the universalizing tendencies of modern technology. In Frampton's view, the fact that we experience the same corporate office blocks in Houston and Helsinki is not an efficient act of international liberation and solidarity, as early moderns had intended, but an act of abstraction and domination by the forces of international capital. This is so because capital is valued in direct proportion to its flexibility, its liquidity, and its ability to be moved from one place to another with minimum resistance. In response to the domination of technology by capital, Frampton has constructed a positive dialectic of the universal and the local.[25] As a modern, Frampton wants to retain the Enlightenment project of universal liberation. However, in order to check the power of capital to abstract and exchange the conditions of daily life, he links human liberation to the realization of the phenomenal conditions of local places.[26]

Frampton's manifesto for a critically regional architecture has appeared in several forms. I will assume the last iteration, which was included in the second edition of his *Modern Architecture: A Critical History* (1985), to be definitive. That essay constructed seven "attitudes," or points, which are indicative of an architecture of critical regionalism. As a reference for my concluding proposal in Chapter 8, I will cite Frampton's seven points in abbreviated form:

1. Critical Regionalism has to be understood as a marginal practice, one which, while it is critical of modernization, nonetheless still refuses to abandon the emancipatory and progressive aspects of the modern architectural legacy.
2. ... Critical Regionalism manifests itself as a consciously bounded architecture, one which rather than emphasizing the building as a freestanding object places the stress on the territory to be established by the structure to be erected on the site....
3. Critical Regionalism favors the realization of architecture as a tectonic fact rather than the reduction of the environment to a series of ill-assorted scenographic episodes.
4. It may be claimed that Critical Regionalism is regional to the degree that it invariably stresses certain site-specific factors, ranging from the topography ... to the varying play of local light across the structure....
5. Critical Regionalism emphasizes the tactile as much as the visual.... It is opposed to the tendency in an age dominated by media to the replacement of experience by information.
6. While opposed to the sentimental simulation of local vernacular, Critical Regionalism will, on occasion, insert reinterpreted vernacular elements as disjunctive episodes within the whole....
7. Critical Regionalism tends to flourish in those cultural interstices which in one way or another are able to escape the optimizing thrust of universal civilization....[27]

On the surface, Frampton's seven points appear to be consistent with, if not inclusive of, all the intentions articulated by the producers of Blueprint Farm.

Those intentions are made explicit in Chapter 4. On that basis, most would deem it appropriate to interpret the selected case as an exemplar of Frampton's theoretical intentions. Such an inquiry would, no doubt, lend support to Frampton's thesis. My own intentions, however, are to allow the inquiry to cut both ways. To simply impose the critical regionalism hypothesis upon Blueprint Farm as a template by which we might interpret a set of objects may make for interesting criticism, but it is an epistemologically questionable enterprise. This is so because the a priori categories of analysis may ignore the assumptions that underlie the project's production. It is a question of methodological *fit*. My intention is, then, to examine the social and material conditions by which Blueprint Farm was constructed in order to gather evidence for the continued renovation of critical regionalism as a theory of architectural production. An axiomatic assumption of this text is that practice should inform theory to the same degree that theory informs practice. A preliminary determination that the objects and the theory to be examined are a good *fit* is only a rationale to proceed, not an understanding of the nature of the fit. Although this inquiry is stimulated by recognizing a correlation between the conditions discovered at Blueprint Farm and those discussed by Frampton, I intend to construct a theoretical position that advances Frampton's hypothesis and, in the process, clears up some problems created by Frampton's eclectic philosophical affiliations.

In his introduction to *Modern Architecture: A Critical History,* Frampton acknowledges his allegiance to the Frankfurt School and a dialectic view of history.[28] Although critical regionalism has emerged from the Marxist assumptions that guide critical theory, I will argue that the practice of critical regionalism, as proposed by Frampton, is inconsistent with those assumptions. Frampton's practice-based departure from critical theory, however, puts him and his supporters at some distance from Theodor Adorno, the progenitor of the Frankfurt School and of critical aesthetic theory. Adorno's position has been one of critically *negative* aesthetic principles. For example, Adorno has argued that any conciliation to the comforting sentiments of local culture only masks the alienation of consumers arising from "that anxiety, that terror, that insight into the catastrophic situation" that characterizes modern life.[29] In this view of negative aesthetic production, the "incomprehensibility" of art is the only means by which the terrible anxiety of our situation might be revealed to us. The incomprehensible is valued by Adorno precisely because it *is* anxiety-producing. To comfort the masses by serving up the local and the familiar is, for Adorno, to participate in the aestheticized politics of late capitalism and the scenographic fantasies of postmodern historicism. Critique— a nihilist philosophical position—is the only social position that is deemed resistant by him to the false comfort of consumption.[30] Adorno finally arrives at an aesthetic position which holds that art can preserve its critical function only by remaining autonomous from the material conditions of everyday life. Adorno's alienation of form from content is relentlessly negative.[31]

In contrast, Frampton's proposal for a critical regionalism rejects nihilism and is finally positive. Although he is highly critical of nostalgia and romantic forms of regionalism, Frampton recommends that we positively (if indirectly) derive architectural forms from the local conditions of place. Where Adorno's position is always aesthetic and psychological in emphasis, Frampton's position leans toward the tectonic and material. The principal difference is that where Adorno hopes for a heightened state of psychological awareness, Frampton demands of architecture a life-enhancing material condition. Whether Frampton intends it or not, critical regionalism opens up a philosophical landscape of radical possibilities that avoid the problematized relationship of modernism to nature. This argument will play a central role in my proposal for regenerative architecture that appears in Chapter 8.

The rapprochement of Enlightenment social goals with a life-enhancing concept of place requires that critical regionalism be associated with progressive, as opposed to conservative, ecologism. Ecologism of any stripe is, of course, considered by most to be a postmodern position because it challenges the modern, anthropocentric worldview. The doctrine of *social ecology*, however, is both rigorously modern and ecologically based. Murray Bookchin is,

gionalism in philosophical discourse, outside the modern/postmodern dichotomy. In the conclusion of this study I will map the *nonmodern* thesis of
this inquiry—one that is, I believe, significant for architectural discourse.

To provide the evidence required to support the nonmodern thesis of
this study I will need to answer two questions, one particular and one general: First, why was Blueprint Demonstration Farm at Laredo, Texas—in spite
of its promise as an experiment in sustainable architectural and agricultural
technology—terminated by the very institutions responsible for its production? Second, how are developments in architectural technology related to
the development of places? The following chapters attempt to answer these
questions.

The fact that I have not included a chapter that sets out the methodological assumptions that I employ in answering these questions requires that
I, at least briefly, comment upon the interdisciplinary nature of this study. While
my subject is principally architecture, I cannot accomplish my mission without help from outside the discipline. I do believe that architecture can claim
ways of knowing that are distinctly its own. It lacks, however, the tactical methods to investigate the ecological and cultural conditions that frame its production. It has been necessary for me to rely upon methods more fully developed in sociology and anthropology to define terms and understand how
architecture participates in the patterns of domination and liberation which
are the concern of moderns. These methods have provided the operational
tools by which architecture might better understand the social construction
of Blueprint Farm in particular and technologies and places in general.

Because I have employed such methods, architects may find literature
referred to in this study to be alien. Architecture, like many other disciplines,
has grown comfortable within a discourse that has become at least self-refer-
ential, if not autonomous. The traditional methods of interpretation employed
in art history, and borrowed by architecture, have assumed that enlightened
observers, like the one portrayed in Figure 1.4, are capable of reading the
object of study directly. Such "reading," of course, depends upon the separa-

Figure 1.4. The *enlightened observer*. From Athasius Kircher, *Ars Magna Lucis et Umbrae* (1671). Courtesy of the Harry Ransom Humanities Research Center, the University of Texas at Austin.

tion of subject and object that is formalized by the social construction of perspectival space that emerged in sixteenth-century Europe.

Some critics add that we also need to consider the intentions of the architects under study if we are to get our interpretation right. Others require that the historian consult the interpretations of other scholars to establish right meaning. The German literary critic Hans Robert Jauss has added to these criteria of art criticism by insisting that we must also understand how the object was socially received in its time in order to get it right. These are all helpful expansions of our understanding of how the interpretation of architecture might be conducted. My own position is, however, that "getting it right" presumes that we can identify the "it" to be investigated. My own definition of the "it" to be pursued is not the architectural object so much as the human practices already incorporated into the object. To understand the politics by which architecture is socially and literally constructed, one cannot rely upon the traditional methods by which art history props up its own authority to interpret and describe at a distance. To distance myself from the fetishism of objects that dominates art-based aesthetic interpretation, I have found it necessary to employ ethnographic methods generally associated with anthropology or sociology. These methods of inquiry get at the social and material conditions which precede the architectural object and which, in turn, derive from the architectural object.

In his *Studies in Tectonic Culture,* Frampton has also relied upon ethnography as a way of knowing architecture. His interest in this field was apparently piqued by the investigations of Gottfried Semper in the mid-nineteenth century. Both architects have examined, 140 years apart, a series of premodern cases in search of the origins of the tectonic—or the poetic—origins of construction. For Frampton, this study has been an effort to capture the "cosmogonic" structure of the Caribbean hut, the Berber house, or the Japanese temple through a study of the available ethnographic literature.[43] In contrast to Frampton's historical use of ethnographic literature, my own study has been an effort to empirically reconstruct the ethnography of a single case. Blueprint Farm was selected as a case study because it is an excellent example of contemporary technological controversy. Rather than attempting to reveal tectonic essences, as Frampton and Semper have done, I have attempted to understand tectonic politics through the interpretation of primary data. Another way to distinguish Frampton's interests from my own is to compare his emphasis upon the "expressive potential" of construction to my own emphasis upon construction as a normative practice.[44] Although these are different interests, they share, I think, many common assumptions.

My intention is that this study will be of interest to those both inside and outside architectural discourse. For an architect to produce knowledge by employing the ethnographic methods of fieldwork is an epistemological challenge to the category of architecture itself. In the end, the interdisciplinary nature of this inquiry may be as important as the problem considered.

In Chapter 2, "A Reconstruction from the File," I tell the official story of the development and demise of the farm through the records which have survived in the archives of the Texas Department of Agriculture, Laredo Junior College, and the Center for Maximum Potential Building Systems. Readers should know that Laredo Junior College "apparently misplaced" large numbers of files related to the controversial closing of the farm, and that the Texas Department of Agriculture was something less than forthcoming in granting access to project files. Access to some documents was, in fact, denied on the grounds that they contained sensitive personnel information. It seems that the existing Republican administration was less than enthusiastic about an investigation into the rapid closure of Blueprint Farm following the 1990 general election. These difficulties in documenting the case serve to amplify the political content of technological innovation and suggest that there may be yet other stories to tell.

In Chapter 3, "The Local History of Space," I reconstruct the agricultural politics that have operated in South Texas since the eighteenth century. With the support of Henri Lefebvre, I argue that the space of la Frontera Chica has been socially produced. The politics of that space helped to elect Jim Hightower, the radical populist Commissioner of Agriculture who initially conceived Blueprint Farm. Hightower and his collaborators had to throw their technological objects into existing political and social situations. Understanding the flavor of the controversy surrounding these objects is, of course, impossible without placing them in the cultural geography of la Frontera Chica. Investigating the history of local space leads me to hypothesize that technologies and places may be different things, but the process of their development is *dialogically,* rather than *dialectically,* related. This chapter concludes with a formal definition of these terms.

In Chapter 4, "Conflicting Intentions," I argue that the dialogue through which technologies and places are socially constructed is shaped by the intentions of competing interests. The chapter documents and relates the competing interests of five distinct groups that were at work in the construction of Blueprint Farm. These competing groups are what the philosopher/sociologist Bruno Latour refers to as "technological networks." Latour's concept and the philosophical literature of "intentionality" lend support to my argument that four out of these five competing technological networks failed to distinguish between the "actual objects" constructed and the "intentional objects" that can be satisfied only by altered human practices.[45] This hypothesis establishes a novel philosophical definition of technological determinism and a demand to reject the subject/object dichotomy that is implicit in modern thought.

Raising the specter of technological determinism requires an investigation of that term and of the technological objects in which so much hope was placed. In Chapter 5, "Technological Interventions," I examine wind-towers, as one example of the sixteen alternative technologies put in place at Blueprint Farm, through the literature of science and technology studies. Although this literature suggests an interpretation of wind-towers as a technology that is more life-enhancing and democratic than the conventional industrial technologies that dominate normative architectural practice, there remain disturbing questions about why such politically positive machines were left in ruin. I conclude that the abandonment of the farm by local farmers is directly related to the fact that they were not engaged in its conception.

In Chapter 6, "Reception," I attempt to understand why local farmers, in whose name Blueprint Farm was produced, failed to adopt the place once the Texas Department of Agriculture and others left the scene. I find that any given cultural paradigm produces both the techniques for interpreting objects and the objects themselves. The extension of this logic is to argue that in order for a culture to successfully receive an alien object, or concept, it must have an already established constellation of ideas to which the imported concept might be grafted. There were as many as six competing paradigms to which the ecologists might have attached their sustainable machines. Tragically, none of these had an established tradition that was receptive to "sustainability" as constructed by the ecologists, nor did the ecologists feel compelled to find a point of attachment.

To have one's claim be reproduced across time and space requires, first, that it be received on fertile ideological ground and, second, that the builders of the claim are able to convince others that their own interests are contained

in its reproduction. In Chapter 7, "Reproduction," I examine how the technological interventions of Blueprint Farm were either reproduced by others, or
disappeared. Although the tracking of sixteen separate interventions is a complex task, one general hypothesis is that the technologies introduced by the
ecologists lacked the "sublimity" desired by locals. Given the choice between
buying trucks—the token and means of mobility along la Frontera Chica—
and buying the organic machines of the ecologists, there was no contest. For
farmworkers living close to the edge, betting on the slowness of natural cycles
seemed to be a very familiar, and painful, idea. Pliny Fisk, however, managed
to construct a sublime reality for his organic machines by importing and converting virtual witnesses via communications technology. There is, however,
an enduring irony in the production of electronic "sustainability."

This study concludes in Chapter 8. There I first summarize the conditions
reconstructed at Blueprint Farm as eight propositions. These provide the
empirical evidence to suggest two alternative theoretical positions by which
the modern/postmodern dichotomy of technology and place might be syn-
thesized—one positive and one negative, but both nonmodern in their assumptions. I briefly discuss the negative valuation of both technology and
place—a position that I term *radical nihilism*—only because it avoids the internal contradiction historically constructed as the modern/postmodern dichotomy between the concepts technology and place. This position is best
articulated by the philosopher Gianni Vattimo and exemplified by the projects
of Rem Koolhaas and the Office for Metropolitan Architecture. Although this
philosophical position warrants further investigation, I find the positive synthesis of the modern/postmodern dichotomy that derives from the doctrine
of critical regionalism to be a more fruitful prospect for inquiry. It is this life-
enhancing alternative that I finally renovate as "Eight Points for Regenerative
Architecture." These propositions provide a practice-based definition of regenerative architecture that supplements the deductive definition provided
in this introduction.

CHAPTER TWO

A RECONSTRUCTION FROM THE FILE

Tax dollars buy new tinkertoys for agribusiness, misery for migrants, death for rural
America, and more taxes for urban America. All in the name of efficiency....

If independent family farmers, consumers, small-town businessmen, farmworkers,
ecologists, farm cooperatives, small-town mayors, taxpayers organizations, labor
unions, big city mayors, rural poverty organizations, and other "outsiders" will go to
the colleges and to the legislatures, changes can occur.

Jim Hightower, Hard Tomatoes, Hard Times, p. 64

This chapter reconstructs the story of Blueprint Farm's production from records
and documents. Where documentation was missing, or just thin, I have amplified that official story by relying upon the recollections of those who were on
the scene. This narrative establishes the facts of the case as those who have
an interest in its telling recorded them. Chapter 3, "A Local History of Space,"
will place these facts in the historical context of la Frontera Chica, and in subsequent chapters I will interpret them.

Jim Hightower became the Texas Commissioner of Agriculture in January
of 1981. In Texas, that position is elective and is, many argue, second in power
only to the governor. Hightower won a bitter campaign against Reagan Brown,

an openly racist "good ol' boy," by appealing to populist sentiments and the promise of agricultural reform. The populist campaign was waged against the assembled forces of agribusiness, the land grant universities, and government itself. Hightower's constituents were an eclectic, but potent, mix of small farmers, ecologists, and businesspeople who shared an interest in economic diversification. In the wake of the antibusiness mood of the late 1970's, Hightower had constructed a network of interests that would last a decade. For progressives, the future looked good—perhaps because it had looked so bad for those outside the entrenched network of business, the universities, and government.

My characterization of Jim Hightower as a populist is hardly unique. He uses this term himself as a description of those who, like himself, wish to more equitably distribute wealth and power to the common people. There is another slant to the historic understanding of populism in the United States, however, that is equally relevant to this discussion. Populism has been a mode of accumulating political power through direct appeal to voters who perceive themselves as estranged. In this sense, populism is not ideological, it is personal. It is not about trusting the historical trajectory of ideas, as is Marxism, but about the historical trajectory of personalities. Before power can be redistributed to the common folk the hero-politician must accumulate it. I intend the term populism to be understood in both dimensions.

Figure 2.1. Jim Hightower as Texas Commissioner of Agriculture. Photograph by Douglas Falls, © 1998 the Jim Hightower Archive, Austin, Texas.

Ascendant politicians take risks. That, and ideological commitment, can be the only rationale for Hightower's early endorsement of Jesse Jackson in the 1984 presidential campaign. Jackson's Rainbow Coalition was an attempt to forge an unprecedented collaboration among marginalized Americans that voiced themes very similar to Hightower's own campaign rhetoric. Shortly after Hightower's endorsement, however, Jackson made his infamous "Hymie" remark, which is quoted by reputable sources and generally accepted as an anti-Semitic slur. That description of Jackson's language is circumstantially relevant to this story because of the repercussions that followed.[1] The immediate political reaction to Jackson's unfortunate remark was to alienate from Jackson his visible supporters within the international Jewish community. Former Congressional Speaker Tip O'Neill is credited with the observation that "all politics are local politics." As if in O'Neill's shadow, Hightower instantly understood that a significant source of his political support, in Texas and around the nation, was about to either desert him or make political life very uncomfortable. The restoration of support from the Jewish community was essential. The question was, how?

In March of 1984, only weeks after Jackson's political blunder, Commissioner Hightower embarked on an official tour of Israel. The official rationale for the trip was simple. "According to him [Hightower], Texas is now facing the water shortages that have plagued Israel since its establishment, and it should *take its cue* from Israel's water conservation pioneers" [emphasis mine]. In Hightower's dramatic and timely construction of common cause with Israel was born a brilliant solution to political (and agricultural) problems. In one stroke, Hightower had wed Jewish interests to a network of small Texas farmers and ecologists, his own political future, and the particular ecological conditions which South Texas and Israel share. On his visit to Israeli kibbutzim, Hightower established relations with Avraham Katz-Oz, the Israeli Deputy Agriculture Minister. Katz-Oz made a similar diplomatic exchange in April of 1985 by coming to Texas. On the basis of these personal ties, plans emerged for a growing number of business and agricultural contacts between the two regions. So successful was this exchange "that a formal agreement—known as the Texas-Israel Exchange (TIE)—was … signed to enable Texans and Israelis to benefit from shared experience and agricultural knowledge."[2]

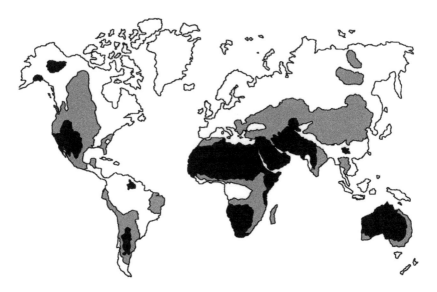

Figure 2.2. Bioregional map. Redrawn from the original by the Center for Maximum Potential Building Systems. The map illustrates the ecological similarities between South Texas and Israel.

As this story unfolds, it will be helpful to remember that Texas was supposed to, in Hightower's words, "take its cue" from Israeli agriculture. The importation by the Texas Department of Agriculture of Israeli technology that was presumed to be superior was the political price that Commissioner Hightower paid for the recuperation of his political viability in the Jewish community. Some understood the Texas-Israel Exchange to be an international agreement for technology cooperation that emerged from the doctrines of bioregionalism. Others, however, understood it to be an international trade concession that emerged from Hightower's bad luck and political instincts for survival.

The Texas-Israel Exchange, or TIE, was conceived as a public-private partnership. Although its offices were physically located within the Texas Department of Agriculture, and its operations received personnel support from TDA, its operational costs and capital projects were financed privately. "The project was started with grants from the Meadows and Hoblitzelle Foundations, the Jewish National Fund, the Texas College and University Coordinating Board, and contributions from corporations, area farmers, and businesspeople."[3] An analysis of the donor organizations indicates that the local Jewish community was primarily responsible for supporting the project. With the control of cash flow comes authority.

As an agricultural model for intensive small farming in a semiarid landscape, the kibbutz is hard to improve upon. TIE soon developed plans to create a kibbutzlike center in Texas that would become a "demonstration and teaching farm to introduce innovative, economically viable sustainable agricultural methods to the region." At the prompting of Dr. Jacinto Juárez, Laredo Junior College (LJC) was selected by the Texas Department of Agriculture (TDA) as the sponsor institution. Although the demonstration farm was initially to be located in West Texas, Dr. Juárez's enthusiasm for the concept convinced Hightower and officials in Laredo to move the project. In the fall of 1987, a temporary site was prepared with a drip irrigation system developed in Israel. The first crop was planted in the spring of 1988. During this planning period, LJC, with TDA support, retained a Laredo architect to prepare preliminary plans for the development of a permanent farm site at the edge of the campus. These plans illustrate a very conventional approach to both space enclosure and technology. It is abundantly clear that these architects, unlike Martin Heidegger, perceived no conceptual link between the act of "edification" and the tilling of soil. Had it not been for the intervention of the Meadows Foundation, it is likely that Blueprint Farm would have been constructed as such a conventional industrial landscape.

The Center for Maximum Potential Building Systems (CMPBS) had previously applied to the Meadows Foundation for funding of sustainable technology research. Pliny Fisk III and Gail Vittori, codirectors of CMPBS, had been developing alternative technological systems for architecture for twenty years, but had been unsuccessful in obtaining such research and development funds because their research had generally lacked a community context, which is a primary criterion for funding by the Meadows Foundation. However, when LJC applied to the Meadows Foundation for funding of the TIE demonstra-

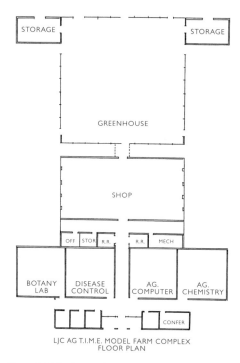

STORAGE

STORAGE

GREENHOUSE

SHOP

OFF | STOR | R.R. | R.R. | MECH

BOTANY LAB | DISEASE CONTROL | AG. COMPUTER | AG. CHEMISTRY

CONFER

LJC AG T.I.M.E. MODEL FARM COMPLEX
FLOOR PLAN

Figure 2.3. The first architectural proposal for the Laredo Demonstration Farm. Redrawn from the original by Cavazos & Associates, Architects, Laredo, Texas.

tion farm, staff members put two and two together. As Alvaro Lacayo recalled it, "Pliny Fisk was married to the Laredo farm by the Meadows Foundation that wanted to give him a grant."[4] At the foundation's insistence, and with TDA's approval, LJC negotiated a contract with CMPBS in December of 1987.

The minutes of early meetings among representatives of the four cooperating organizations reveal enthusiasm for a radical rethinking of all previous architectural concepts. It must have been obvious to these participants that the reason for their enthusiasm was the fact that the categorical boundaries between agriculture and architecture were being erased by the committee's decisions. As a student of both architecture and landscape architecture at the University of Pennsylvania in the era of Louis Kahn and Ian McHarg, Fisk had long ago lost track of the disciplinary boundaries between architecture and ecology.[5] His charismatic self-assurance seemed to stretch his collaborators beyond their experience. The documentation of this discourse suggests a heady and infectious experience that produced very high expectations.

As the first experimental crop tentatively emerged from the field, planning for the permanent site emerged from intense research and discussion. But, by April of 1988, with construction documents due on June 20th, frustrations were growing as well. Fisk found it necessary to admit that "I didn't realize the amount of work required to get everything together in a coordinated manner."[6] He was also not accustomed to working in a bureaucratized institutional context that demanded the degree of accounting minutiae expected by TDA. Final construction documents were once again promised for June 20th, as originally scheduled.

The elemental level at which the participants were assessing technology was a formidable task. This was not a process of assembling off-the-shelf industrial components born of an energy-consumptive society. The very idea of roof, or wall, was being reinvented as one element of an ecological system. The task defined by these collaborators was to construct a technology that united means and ends, process and product. A scarcity of water in one location was not to be solved by a thoughtless expenditure of other, equally valuable resources mined at some distant location. Fisk was taking the TIE charter mandate to implement "sustainable agricultural methods" quite seriously. By September, the evaluation of alternative technologies was still very much in progress. Fisk, in a review of pending decisions, wrote to his collaborators that "I am taking this opportunity to address some of the specific technologies, but, more important, to show how technology choices can help to establish an overall farm policy; that is, what is this farm all about[?]"[7] For Fisk, it was self-evident that technology is not merely instrumental means. The way that the farm would be built was, in his view, the agricultural policy. However, his repeated attempts to get policy makers interested in the social construction of elemental technologies, like natural cooling, for example, were beginning to wear thin. Production was beginning to fall behind schedule, and TDA was

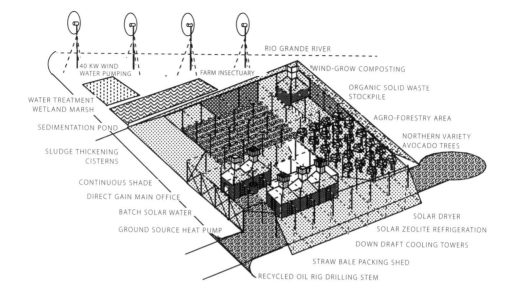

RIO GRANDE RIVER

40 KW WIND
WATER PUMPING
FARM INSECTUARY
WIND-GROW COMPOSTING

ORGANIC SOLID WASTE
STOCKPILE

WATER TREATMENT
WETLAND MARSH

AGRO-FORESTRY AREA

SEDIMENTATION POND

NORTHERN VARIETY
AVOCADO TREES

SLUDGE THICKENING
CISTERNS

CONTINUOUS SHADE

DIRECT GAIN MAIN OFFICE

BATCH SOLAR WATER

SOLAR DRYER

GROUND SOURCE HEAT PUMP

SOLAR ZEOLITE REFRIGERATION

DOWN DRAFT COOLING TOWERS

STRAW BALE PACKING SHED

RECYCLED OIL RIG DRILLING STEM

Figure 2.4. The original CMPBS proposal for Blueprint Farm, by Pliny Fisk III, © 1988 CMPBS. Courtesy of the Center for Maximum Potential Building Systems.

experiencing political pressure to produce a visible ideological success as well as the artifacts themselves. Only six months after those enthusiastic conceptual discussions, competing interests were beginning to emerge.

As the CMPBS staff experimented with alchemical formulas of fly-ash, straw, and recycled steel, TDA staffers in Austin experimented with political alchemy. In a March draft of the "Mission Statement for the Laredo Blueprint Demonstration Farm," a TDA official named John marked up a copy to require the writer to stress three components more forcefully in the next draft: "(1) sustainable agriculture, (2) ties with Israel and eventually with Mexico, and (3) assistance to small acreage producers."[8] A more transparent, if not cynical, way to list these categories would be, not as missions, but as votes. They translate as (1) ecologists, (2) Jews and Hispanics, and (3) small Texas farmers. Of course, a generous reading of the revised mission statement suggests that TDA was attempting to involve the broadest possible constituency in tech-

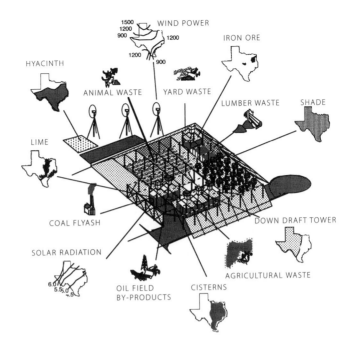

WIND POWER
1500
1200
900
1200
1200
900
IRON ORE
HYACINTH
ANIMAL WASTE   YARD WASTE
LUMBER WASTE   SHADE
LIME
DOWN DRAFT TOWER
COAL FLYASH
SOLAR RADIATION
AGRICULTURAL WASTE
6.0
5.5
5.0
4.5
OIL FIELD   CISTERNS
BY-PRODUCTS

Figure 2.5. Details of the CMPBS proposal for Blueprint Farm, by Pliny Fisk III, © 1988 CMPBS. Courtesy of the Center for Maximum Potential Building Systems. The figure illustrates the incorporation of regional materials into the project.

nological innovations that would affect not just farmers, but the economy and the environment as a whole. Both interpretations are warranted.

As if to rationalize the project's emerging financial woes, Alvaro Lacayo reflected that "Radical change takes a lot of money."[9] As the mission was being refined over the next nine months of research, development, and construction, unanticipated costs were creeping in. March of 1989 also brought the end of Phase 1 funding by the Meadows Foundation and the application for Phase 2. Being $37,000 over budget did not seem surprising to anyone familiar with the scope and experimental nature of the project. But it did create tension and threaten CMPBS's credibility with an account-conscious public agency eager to boast about economy. Discussing decisions about ceremonial moments, like who would cut the ribbon on May 12, 1989, brings conflicting versions of reality to the surface. Fundamental questions of authority, such as whose name will appear on the entrance sign, must be asked,

Figure 2.6. Jim Hightower and visiting dignitaries inspecting crops at Blueprint Farm. Photograph by Karen Dickey for Jim Hightower © 1998 (?).

and answered. In this case, Fisk did not like the answer he received about the sign, so he portrayed reality as he saw it. According to the TIE coordinator, Fisk was responsible for getting the sign painted, but neither his name, nor that of his organization, should appear. By bureaucratic decree, Fisk, Vittori, and their associates were rendered officially invisible in the "exchange" with the Israelis. For the Department of Agriculture, the presence of architecture must have been deemed confusing to the agricultural interests being assembled. It was, therefore, suppressed.[10]

A few days after the grand event attended by dignitaries from Israel, Mexico, and all corners of Texas, Vittori received an outraged memo from the same TIE coordinator. She advised Vittori indignantly that she had directed the sign painter to change the text of the sign back to that drafted by TDA. This comic exchange, of course, represented a tragic confrontation over authorship. Whose project was this anyway? Without warning, the first salvo of an answer came the next day. CMPBS was formally advised that funding for the "building project" would now be separated from the "farm project." It was the competition for public signature that precipitated the economic DMZ that separated farming and building at Blueprint Farm. As briefly discussed in Chapter 1, the significance of this administrative categorization cannot be overemphasized.

From TDA's perspective, an "internal assessment of the Laredo situation" was long overdue. In the internal memorandum that documented the review of problems in Laredo, four major problems were identified by TDA:[11]

1. "Overall, clarity is missing as to the delineation of who is in charge and who has the final word...."
2. "...there is a serious and clear absence of organizing at the local level." The TDA assessment document indicates that Laredo Junior College had made virtually no effort to engage the local community in the project. If TDA officials in Austin recognized the absence of local participation to be a problem, the seriousness of this omission cannot be overemphasized.
3. LJC was also characterized in the assessment document as being relentlessly bureaucratic in the administration of farm equipment and funds—a situation that was incompatible with the very nature of farming. And finally,
4. The assessment document also reflected the cultural conflict between the Israeli farm manager and his adopted place. The ecologists later characterized this conflict as more ideological than cultural.

Following the TDA review and subsequent economic partition of "the farm project" and "the building project," Fisk and his ecologist collaborators were pushed further and further into the margin of the "exchange" with Israel. In late 1989 the official and unofficial project files became a repository of increasingly bitter missives. In frustration, Fisk concluded that it was time to reveal the heavy use of pesticides, herbicides, and fungicides by the Israelis on their so-called "sustainable" farm. In correspondence to TDA he demanded to know if "sustainable practices on the farm [are] other than to say we are saving water?"[12] Fisk did make these accusations public. Months later, in March of 1990, TDA responded by issuing a brief statement that "pesticide use requirements have been followed very carefully" at the farm.[13] There was no technical disclosure as to how TDA reached this conclusion—case closed.

But, while these investigations were ongoing, the administrative reorganization of the project began to take shape. In December 1990 management of the project was transferred from Laredo Junior College to Texas A&I University, although LJC remained the official host. CMPBS was left to negotiate a new relationship to the project with Dean Donald Hegwood of TA&I. Dean Hegwood's first correspondence with Fisk was to deny his proposal to continue the composting project on the site. The continuing allegiance to what Fisk perceived to be Israeli interests only made him feel more isolated and betrayed by the Hightower regime. By this time, Fisk's rage could barely be contained. He penned a poisonous letter to TDA, which was never mailed, that asked very sensitive questions about the political motivations of Hightower's policy toward technology and toward Israel and its technology in particular. A reader of that file can only wonder at what might have transpired if that letter had been mailed.[14]

Acrimony and accusation from both sides of the DMZ became commonplace in 1990. Fisk initiated contacts with new sources of funding, only to be reminded by TDA that he had no authority to make commitments for the property without the approval of TDA, TA&I, and TIE. By July of 1990, the TIE coordinator voiced concerns about the construction completion schedule and items "scheduled as completed that still don't work successfully." Money was again running low, and the site was, she claimed, "a mess." The TIE coordinator obtusely asked, ". . . can we count on the fact that your products will be completed and working properly? And, if so, when?"[15] The reference to "products" by the TIE coordinator is helpful in understanding how TDA had come to categorize CMPBS's work. Architects normally provide services because they do not participate in making the thing itself. Contractors build buildings. Farmers produce crops. It is manufacturers, however, who make "products." These products are commodities, like air conditioners, that satisfy discrete consumer demands. CMPBS performed all of these roles—designer, builder, farmer, and manufacturer—but was conceptually and operationally contained by none of them. In her frustration to categorize CMPBS, the TIE coordinator imposed that category which would best satisfy her immediate needs. That limiting definition was increasingly unsatisfactory to Vittori and Fisk.

It will be helpful in this situation to introduce the distinction made by Henri Lefebvre between a "work" and a "product." In Lefebvre's view, a "work" is a thing that has something irreplaceable and unique about it. A "product," on the other hand, is a thing that can be reproduced exactly. Using Lefebvre's terms, CMPBS intended to produce a "work"—a set of spaces and technologies that were uniquely tuned to conditions at Laredo. TDA, it seems, expected to receive a "product"—a set of appliances that could be infinitely reproduced.[16]

In spite of the firefight across the DMZ that separated the two projects, in September 1990, Fisk was appointed to the newly formed LJC Demonstration Farm Steering Committee. How might we understand this fact? As usual, money is an answer. Although his institutional collaborators had effectively marginalized Fisk, the Meadows Foundation, which believed deeply in what this coalition of ecologists was trying to accomplish, still held the purse strings. This was one relationship that, as Fisk put it, never "went weird."[17] So long as Meadows continued to back CMPBS's view of a sustainable world, TDA would have to tolerate its presence. Although the project was nearly closed down in the summer of 1990 for lack of funds, the Meadows Foundation opened its coffers once more to keep the "demonstration" alive.

Figure 2.7. Benjamin Gamliel, the first Israeli farm manager of Blueprint Farm. Photograph by R. Norman Matheny, © 1989 *The Christian Science Monitor.*

If any reader has been able to keep track of the political timeline of this saga, he or she will have noticed that the electoral clock has been ticking steadily. To Jim Hightower's great surprise, he was not reelected in November of 1990. The regime of sustainability came to an end as quickly as it had emerged. In early January 1991, Dean Hegwood corresponded with the new assistant agriculture commissioner responsible for the project, advising him of pending grant proposals.[18] By May 1991, the Meadows Foundation had signed off on the now fulfilled grant responsibilities of CMPBS. Although the project did not yet operate, as designed, the Meadows Foundation had seen enough. A few days later, the new assistant commissioner wrote to the Hitachi Foundation—grantor of the only unexpended funds—to request that the balance be used "to enable Texas farmers from more than one area of the State to adopt the use of Israeli technology."[19] These events made it clear to all the participants that, despite claims to the contrary, the new Republican admin-

istration intended to terminate this problematic political legacy. It was over. Jim Hightower, recalling the political origins of TIE (unwittingly unleashed by Jesse Jackson's imprudent remark), must have found the consequent redirection of the project funds ironic. The new Republican administration headed by Commissioner Rick Perry proposed to make the remaining funds available to African American agriculture students from Prairie View University so that they might travel to Israel.

In her transition memo to her Republican successor, the TIE coordinator, who had become Fisk's nemesis, described him with painful frustration:

Figure 2.8. Rick Perry as Texas Commissioner of Agriculture. Perry succeeded Jim Hightower and presided over the closure of Blueprint Demonstration Farm. Photograph courtesy of Texans for Rick Perry.

First, he doesn't seem to ever get done (and doesn't finish according to agreed upon timetables); second, he cannot seem to finish anything within the budget (twice he has had to be bailed out with extra money from Meadows); thirdly, he cannot stay focused on the tasks that are his responsibility without straying into every other area of the farm and deciding that he has to play an integral role in everything; fourthly, when things didn't go his way, he bad-mouthed everybody and then denied it. He sees the world in black and white (anyone who disagrees with him and challenges his grandiosity becomes his enemy). He is a man of ideas who should never be given the responsibility to implement them.[20]

Such a stinging condemnation is evidence on the side of those who desperately wanted to complete a project. From inside the well-ordered boundaries of conventional political interests, Fisk's deconstruction of the boundaries between agriculture and architecture, between means and ends, between art and science, must have seemed more than eccentric. In the eyes of those who valued the efficient production of reliable products, his behavior was unprofessional—meaning that Fisk did not "profess" the same categories of knowledge and interpretation as did his institutional collaborators.

From the other side of the DMZ, Fisk constructed for himself a conspiracy that converged on his competition with Israeli interests. He perceived that on the outside of the axis of power were unappreciated Texans, like himself and his partner, Gail Vittori, who were expected to remain anonymous in the face of Jim Hightower's political aspirations. Those aspirations depended, of course, upon the presumption of Israeli technological superiority. Having invested the previous twenty years of his life in the research of appropriate, regional technologies, Fisk found it impossible to imagine that the concept of sustainability might be employed for symbolic or political ends. When ideology was betrayed, it was sadly necessary for him to explain the erosion of his own interests in conspiratorial terms. Ideology begets ideology.

The story of Blueprint Farm did not end in 1991. Although the site lay

Figure 2.9. Pliny Fisk III and Gail Vittori, codirectors of the Center for Maximum Potential Building Systems, Austin, Texas. Photograph by Paul Bardagjy © 1998.

deserted for a few years, attaining the patina of an archaic ruin, the deteriorating buildings were leased in September 1994 by the Rio Grande International Study Center (RISC), a nonprofit river-study, river-watch organization funded by LJC, Webb County, the City of Laredo, and various grants. The ecological agenda of this organization is certainly sympathetic to the initial purpose of the demonstration farm, but the site and buildings were hardly suited to RISC's administrative and pedagogical purposes. Although the "machines" that Fisk designed for agricultural purposes are no longer in use, they continue to be admired for their aesthetic presence. A project to permanently renovate the site as an interpretive "museum," or "living lab," for the Rio Grande International Study Center was conceived in 1995. The conversion of the site was designed by Sepúlveda Associates Architects of Laredo, and construction was completed in 1999. Shortly after her appointment, the first director of RISC wrote to Fisk expressing her appreciation of "how the buildings look, how they feel, and the spirit and effort that went into the concept and construction of the site."[21] After so much controversy over the technological operation of the farm, Fisk was pleased to accept aesthetic praise.

CHAPTER THREE

THE LOCAL HISTORY OF SPACE

If space is produced, if there is a productive process, then we are dealing with history;...Since, *ex hypothesi,* each mode of production has its own particular space, the shift from one to another must entail the production of a new space.

Henri Lefebvre, *The Production of Space,* p. 46

...in the case of technological networks, we have no difficulty in reconciling their local aspect and their global dimension. They are composed of particular places, aligned by a series of branchings that cross other places and require other branchings in order to spread.   Bruno Latour, *Science in Action,* p. 131

The politics inscribed in the objects and social spaces of Blueprint Farm did not originate with Jim Hightower's plan to revolutionize Texas agriculture. Nor did they begin with Pliny Fisk's notion of "flexible farming." Nor with Israeli kibbutzim. To attempt an understanding of Blueprint Farm within the limits of recent events would be to suppress the cumulative history of space. The producers of Blueprint Farm did not invent their situation. Rather, they contributed to a continuum of events that have piled up in la Frontera Chica, or "the little border," as the inhabitants of Zapata, Starr, and Webb Counties refer to their region. To reconstruct how locals understand the historical continuum of their own practices will require some excavation.

This chapter is in three parts: First, it is necessary to define what I mean by the terms *place* and *technology* and to generally relate these constructs to each other. Second, these definitions provide a theoretical context for constructing what Henri Lefebvre would call a "local history of space." In Lefebvre's view, "A space is not a thing, but a set of relations between things."[1] Or, put even more directly, "(Social) space is a (social) product."[2] The second goal of this chapter, then, is to reconstruct the production of la Frontera Chica as a social morphology. This will be derived from interviews with locals and from local histories. Third, I interpret that history of space through the abstract definitions of place and technology constructed in part one. The chapter then concludes with an empirically derived hypothesis that relates the concepts of place and technology.

PLACE, TECHNOLOGY, AND TECHNOLOGICAL NETWORKS   My definitions of place and technology make use of arguments developed in the disciplines of cultural geography and sociology. From cultural geography my analysis relies most directly upon the works of John Agnew and Henri Lefebvre because they have been particularly influential in resurrecting the concept of place as a subject of serious study. I should be quick to point out that Lefebvre's discourse concerns the social construction of "space," not "place." As I pointed out in Chapter 1, these concepts are historically distinct, and occasionally allergic to each other. In this chapter, however, I will use them interchangeably because Lefebvre's project is itself a mixing of modern and postmodern, Marxist and Heideggerian, ecological and political assumptions. Neil Leach, for example, categorizes Lefebvre as a phenomenologist where others categorize him as a Marxist.[3] Such confusion about his assumptions makes Lefebvre the ideal precursor to a nonmodern thesis. From sociology my analysis relies upon texts by Donald MacKenzie, Judith Wajcman, and Bruno

Latour.[4] Latour, in particular, has contributed terms to the critique of modern technology that are spatial, and thus related to the concerns of architecture.[5]

As a cultural geographer John Agnew argues that places cannot be understood within the limits of architecture or physical geography. There is a growing body of literature, originating with Heidegger, which meditates upon the question of boundaries.[6] These authors ask, where do places begin and end, and with what senses do we find them? Agnew's answer is that the qualities of place are complex, quantitatively and qualitatively. He offers three qualities through which we might understand the phenomenon of place: *location, locale,* and *sense of place.*[7] I will use these concepts as lenses through which to study la Frontera Chica.

By location, Agnew intends that a place can be understood as "the geographic area encompassing the settings for social interaction as defined by social and economic processes." This quality of place includes the objective structures of politics and economy that link one place to another: the EC (European Community) and the Monroe Doctrine are examples derived from formal political alliances and economic structures. The alliances of corporate economies also construct locations. Houston, for example, is effectively closer to the east coast of Scotland than to Arkansas because these oil-producing landscapes are managed by the same corporate structures. It is these structural conditions of political economy that most concern Marxist scholars.

If location is defined as a set of objective structures, Agnew argues for the existence of a sense of place as a set of intersubjective phenomena. By this term he means the local "structure of feeling" that pervades Being in a particular place. This quality of place includes the intersubjective realities that give a place what conventional language would describe as *character* or *quality of life.* For example, the reverence that the citizens of Austin, Texas, reserve for a swim in Barton Springs or the stylish ambition of street life that New Yorkers enjoy are ontological, rather than physical, qualities of place. It is in this mode that the complex human poetics of place are experienced. It is also in this mode that constructivist scholars study the intersubjective construction of reality.

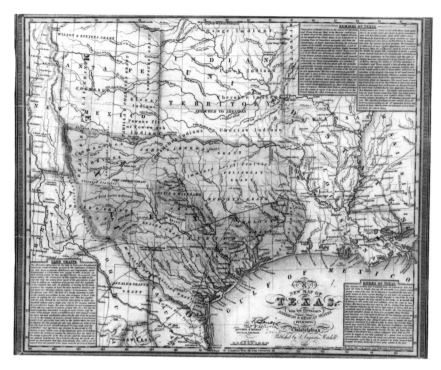

Figure 3.1. A pre-1845 map of Texas. Note that the area between the Rio Grande and the Nueces River was then part of Mexico. Courtesy the Texas Map Collection, the Center for American History, the University of Texas at Austin, CN06614.

Between the objective quality of location and the subjective quality of sense of place, Agnew establishes a middle ground, or locale. This quality of place is the setting in which social relations are constituted. Locale includes the institutional scale of living to which architecture contributes so much: the public square, the block, and the neighborhood. I have chosen this topos, or philosophical place, from which to observe Laredo in the third section of this chapter.[8] From this quasi-objective, quasi-subjective place I attempt to align the structures of location that hover above us with the sense of place that we experience on the ground. My intent in this operation is to avoid the overdetermination that derives from a Marxist preoccupation with the conditions of political economy and at the same time avoid the underdetermination that derives from a constructivist preoccupation with the conditions of atomized reality.[9]

Figure 3.2. A typical street scene in Laredo. On the U.S. side of the border, outlets for consumer tourism dominate the landscape. Author's photograph.

I should stress here that Agnew's distinction among location, locale, and sense of place is not simply a matter of macro-, meso-, and micro-scaled analysis. Rather, it is the "elastic" interaction among all three qualities that constitutes place in Agnew's terms. "It is the paths and projects of everyday life, to use the language of time-geography, that provide the practical glue for place in these three senses."[10]

If the concept of place requires such a multifaceted definition, what about technology? Just as conventional thought understands place as only physical in quality, technology is commonly understood as physical hardware. Such a physicalist definition tends to consider the social construction of automobiles or refrigerators, for example, as outside the competing interests of society.[11] In the tradition of positivism, technology is understood as the asocial application of scientific truths. In contrast, the literature of science and technology studies—which is discussed more thoroughly in Chapter 5—has demonstrated that technology, far from being constructed outside society, is a system that is inextricably part of society. Technology, like place, is a field where the struggle among competing interests *takes place*. MacKenzie and Wajcman have argued that the concept of technology, like place, includes three qualities. In their construction, technology includes *human knowledge, patterns of human activities,* and *sets of physical objects.*[12] I will also employ these concepts as a second set of lenses through which to study la Frontera Chica.

In MacKenzie and Wajcman's definition, knowledge—the first characteristic of technology—is required, not only to build the artifact, but also to relate the natural conditions upon which the artifact works, and to use the artifact. The second characteristic of technology, "patterns of human activity," or what I would prefer to call *human practices,* refers to the institutionalization, or routinization, of problem-solving that inevitably occurs in society. The practices of architecture, carpentry, or farming are examples. The third quality of

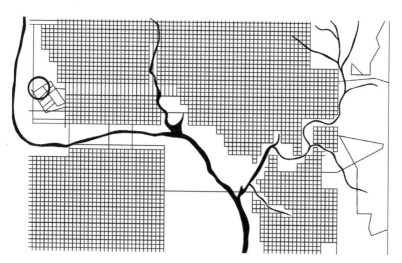

Figure 3.3. Map of downtown Laredo and Nuevo Laredo. The site of Blueprint Farm was on the grounds of Laredo Junior College at the left of the map between the international railroad and the bend in the river. Redrawn from Rand-McNally, Map of Laredo.

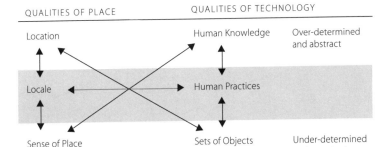

Figure 3.4. The dialogic qualities of place and technology.

technology, "sets of objects," is, of course, the most obvious—these are the things themselves. The point is, however, that computers, hammers, or tractors are useless without the human knowledge and practices that engage them.

What I want to argue here is that the definition of place offered by Agnew, and the definition of technology offered by MacKenzie and Wajcman, are related by a tripartite structure that is not accidental. Figure 3.4 will help to make this point clear.

The limited point of the diagram is threefold: First, that places and technologies are both spatial concepts with related structures. Second, that these qualities are dialogically related. And third, that modern forms of knowledge, like the economics of location, tend toward the abstract and overdetermined, while our understanding of objects and sense of place tends toward the underdetermined and the atomized. These points serve only to magnify the centrality of locale and practices as the glue that holds the discourse of places and technologies together.

To argue that place is a spatial concept is a tautology and requires no further backing. However, to argue that technology is a spatial concept requires some explanation. Bruno Latour, to whom I briefly refer in Chapter 1,

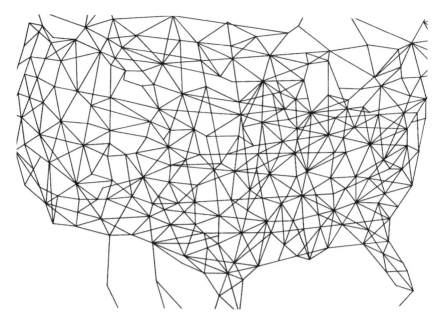

Figure 3.5. Highway linkages in the United States. Highways are, as Bruno Latour suggests, a literal net thrown over space. Redrawn from AAA road map of the United States.

has argued that "Technological networks, as the name indicates, are nets thrown over spaces."[13] By "technological network," Latour refers, not just to "sets of objects," but to the social networks that construct a relationship among human knowledge, human practices, and nonhuman resources—the stuff from which the objects themselves are made. His point is that technology is essentially a spatial concept because its operation depends upon the mobilization of human and nonhuman resources *that exist in different places*.[14] For example, architects, clients, contractors, and bankers comprise a social network of building producers. Their relationship has a social and spatial quality to it. Advances in communications technology, however, have radically collapsed the spatial reality of these social relationships. When one recognizes, however, that lumber from Oregon, windows from Pittsburgh, carpet from Mobile, and compressors from Taiwan are required to realize the material intentions of the producers, the concrete qualities of their purely *social* network are materialized as a global *technological* network. A technological net-

Figure 3.6. Truck lot below the international bridge at Laredo. Author's photograph.

work produces spatial links that tie the social network of producers to those nonhuman resources required for construction. This is a central argument of this study that has, as we shall see shortly, important implications for the social construction of la Frontera Chica as a place.

To follow Latour's argument and the relationship of technology and place constructed in Figure 3.4 leads to a central argument of this chapter. This proposes that technology is best understood not through history, but through geography. History tends to interpret reality as *human events in time*. Through temporal interpretation we might better understand the causal sequence in which humans construct artifacts. In contrast, geography tends to interpret reality as *human events in space*. Through spatial interpretation we are more likely to understand how technological networks operate to dominate the places inhabited by humans and nonhumans. It is geography, then, that offers methods more relevant to this inquiry. It is in the local history of space that the relationship of places and technologies becomes concrete.

Henri Lefebvre has argued two points that reinforce the dynamic relationship between technology and place that is claimed here. First, that places are produced by technology acting upon nature. Implicit in this point is the claim that original nature, if it ever existed at all, has long ago been incorporated into *second nature,* which is a work of society.[15] Lefebvre's second point is that each society—or, as Marxists would have it, each mode of production—produces its own peculiar type of space.[16] I have cited this point above, but in this context Lefebvre's argument serves to explain that the differing qualities of places are more a matter of technological practices than aesthetic choices.

In constructing this dialogic relationship between place and technology, I should make clear that I am not building a case for environmental determinism—places do not *cause* technologies. Given different cultural conditions, the sets of objects that dominate a place like Laredo might be different. Given constant environmental conditions, the interpretive flexibility of culture is entirely contingent. I will argue that environments do shape technologies, but are in turn shaped by them.[17] As a corollary, I am not building a case for

Figure 3.7. The ruins of Guerrero Viejo. The old village of Guerrero was flooded by the construction of the Falcon Dam. The sustained drought of the 1990's, however, allowed the old village to reemerge. Author's photograph.

technological determinism—technologies do not *cause* places. The same logic holds that technologies *do* shape places, but are also shaped by them.[18] The important question of technological determinism will be taken up again in Chapters 4 and 5. The point here is that the relationship of place and technology is both spatial and discursive. It is a dialogue of cause and effect, means and ends. They are inseparable, but contingent, concepts that lead inhabitants of a place to a dialogic narrowing of cultural horizons. This abstract proposition, however, requires concrete evidence to become relevant to our discussion.

This section provides the empirical evidence to support the dialogic definitions of technology and place proposed above—it constructs a local history of space that is told principally by locals through interviews and written histories. These texts tell us that, unlike San Antonio to the north or Goliad to the northeast, Laredo was never the site of a colonial Spanish mission.[19] The abbreviated story goes something like this:

LA FRONTERA CHICA

In the mid-eighteenth century, Don José de Escandón, the impresario who established Nuevo Santander (the Spanish colony that is roughly congruent with the Mexican state of Tamaulipas and Texas south of the Nueces River), sent his aide, Thomas Sánchez, to found the new town of Laredo. In 1755, after several failed attempts to find a suitable site on the Nueces River to the north, Sánchez and his *pobladores*—those colonists who would populate the Spanish Empire—settled on a site on the north bank of the Rio Grande del Norte where two roads converged. The first went to the mission at San Antonio, the second to La Bahia (Goliad). This site had been located earlier by soldiers who forded the river at a location known as Paseo de Jacinto. Even though Laredo was not a colonial mission, it was settled in accordance with the Laws of the Indies, which first appeared in 1573 and subsequently became codified as the Spanish Orders for Discovery and Settlement.[20] This or-

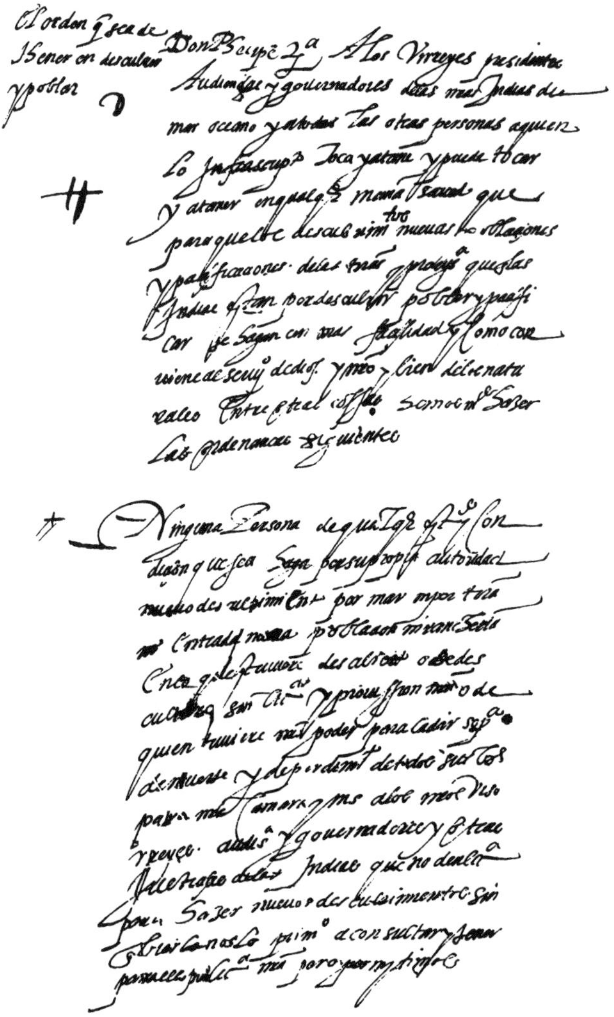

Figure 3.8. The Spanish Orders for Discovery and Settlement of 1573. In Spanish, Ordenanzas de descubrimiento, nueva población y pacificación de las indias dadas por Felipe II, el 13 de julio de 1573, Archivo de los Indios Sevilla. The original of this document is located at the Archives of the Indies at Seville, Spain. This text describes how to construct a colonial Spanish city in terms that we now recognize as conceptually Cartesian, in that they are based upon a priori abstractions rather than upon local conditions. For example, article 114 of the ordinance reads: "From the (main) square there should start four main streets in the middle of each side and two streets by each corner of the square. The four corners of the square should face the four main winds because in this way the four main streets coming out of the square will not be exposed to the four winds, which would be very inconvenient." Translation by Jan García.

Figure 3.9. The pattern of *porciones,* or agricultural lands, distributed along the Rio Grande by the Spanish crown in 1767. From the date of settlement in 1755 to the General Visita of 1767, lands were held in common, or in a state of *ejido.* The linear pattern of agricultural lands along the river was to ensure water access. Redrawn from the original by Jack Jackson, *Mapping Texas and the Gulf Coast: The Contributions of Saint-Denis, Olivan, and Le Maire* (College Station: Texas A&M University Press, 1990).

dinance required a rigorously orthogonal and hierarchical layout of the residential and commercial sectors of the town—a purely Cartesian geography—and a ribbonlike distribution of agricultural lands along the river—a geography responsive to local ecology in its assurance of access to water.

The fact that Laredo was not a mission suggests that it sprung up, not as a place to Europeanize indigenous peoples, but as a place to strategically resettle Europeans already in Mexico. Although Laredo marked a strategic crossing on the Rio Grande, it is upriver from the prime agricultural lands of the Lower Rio Grande Valley, and very far downriver from prime agricultural lands located near the headwaters of the Rio Grande in what is now New Mexico. Other than in the thin strip of agricultural land along the Rio Grande, agricul-

Figure 3.10. Typical range scene in la Frontera Chica. Author's photograph.

Figure 3.11. Industrial farming in la Frontera Chica. Author's photograph.

tural practices were a marginal prospect for the citizens of Laredo. Intensive farming still exists around the Rio Grande river delta one hundred miles downriver, and in New Mexico several hundred miles upriver. However, the "bad to horrible" quality of the local soil virtually ensured that agriculture would not play a dominant role in the economy of la Frontera Chica.[21] Although cash crops of onions and melons contributed to the local economy well into the twentieth century, the agrarian history of Laredo has been linked more to a colonial, hierarchical model of ranching than to the pattern of pastoral subsistence that is common among the Hispanic settlements of New Mexico.

The marginal farming that has survived in la Frontera Chica is, however, in a period of fundamental transition. Although the total volume of agricultural production is in decline, those lands still in production have become economically consolidated and chemically dependent.[22] Economic consolidation stems from two conditions: First, the absence of a Hispanic tradition of primogeniture, or the inheritance of whole agricultural properties by the eldest heir, and second, the imbalance of wealth in the region. Without a tradition of primogeniture, agricultural properties have been subdivided into successively smaller and less productive holdings as generations have succeeded each other. In order to survive, later generations have been forced to sell what little capital they have inherited. These economic conditions conspired against the continuity of pastoral practices. Those with available capital were in a unique position to consolidate and mechanize agricultural production. As Jim Hightower characterized the situation for his constituency:

> The technology of agricultural mechanization has altered more than the labor input of planting, thinning and harvesting. It has also meant the genetic alteration and chemical treatment of crops to render them more adaptable to mechanical harvesting.[23]

Local farmers understand quite clearly that the radical mechanization of agricultural production in the United States, Mexico, and Central America has displaced labor and altered rural life worlds beyond recognition.[24] The majority of the displaced farmworkers in la Frontera Chica are, in fact, not from there. Rather, they have been displaced from more agricultural communities to the south and have come to Los Dos Laredos to seek their fortune in the relative prosperity of the border region. As often as not, however, they find marginal work that includes agriculture. As Roberto Elizondo understands it, those laborers who can find work are paid much better these days, but they "won't work outside of an air-conditioned cab."[25] Hightower theorized such local insights by arguing that

The beneficiaries of agricultural modernization are first the large-scale growers capable of the massive capital investment required to gain access to technology, and second the industrial producers of the machines. Both groups are direct recipients of land grant complex support, in contrast to the small farmers and laborers who have been increasingly excluded and marginalized.[26]

In Hightower's view, technology has operated as the agent of those who would concentrate production in the hands of a few, and as the enemy of those who would liberate themselves from bondage to the land. His project was to invert the institutional systems that support agribusiness and to invert the agency of technology.[27] Hightower dared to imagine appropriate technologies accessible to displaced farmworkers in lieu of capital-intensive technologies accessible only to large landowners.

Before construction of the Falcon Dam (about sixty miles downriver from Laredo) in 1950, most farmers in la Frontera Chica operated "dry" and depended upon the whims of nature to supply water. But now, because of the reliability of the Falcon Reservoir as an irrigation source, most commercially viable farms have switched to subterranean gated pipe irrigation systems. Consolidated farms irrigate as much as 4,000 acres.

Because of the recent professionalization of farm practices in la Frontera Chica, most prominent Laredoans assume that small-scale farming is a dying practice. They rationalize this prediction on the basis of prolonged drought in the 1990's and what farmer Roberto Elizondo called the "inevitability of progress."[28] Elizondo's view conflates space with time; sooner or later, the inevitable trajectory of technological modernization reaches even the most

Figure 3.12. Downstream view of the Falcon Dam powerhouse, constructed in 1950. Author's photograph.

Figure 3.13. A typical oil well in Webb County, Texas. Author's photograph.

remote places. Most contemporary Laredoans assume that farming was never a significant local practice in the first place. For upwardly mobile Laredoans, farming is perceived not as the productive vocation of peasants, but as a consumptive avocation of the rich. Large Webb County landowners generally describe themselves as "cattlemen." Landless workers, however, describe these elites as "oil men with an expensive hobby."[29] Hightower the populist would, no doubt, enjoy such a characterization. He didn't enjoy, however, the historical determinism implicit in local attitudes toward the inevitability of technological modernization. To those who listened, the attitude he advocated was darkly ironic. Although the evidence indicates that agricultural production per worker has increased dramatically, the consequences for workers themselves have been disastrous. As early as 1978, before his election as Texas Agriculture Commissioner, Hightower had complained that

In 1940 a farmer could feed himself and eleven others. By 1970, that ratio had grown to one to forty-five. Americans now spend only 16% of their disposable income on food—the lowest rate in the world. These numbers, of course, do not account for the deferred social and environmental costs of rationalized production.[30]

Alvaro Lacayo, a local farm activist, is a Hightower supporter and only too aware of the "deferred social and environmental costs of rationalized production." He prominently displays twin portraits of Che Guevara and Willie Nelson above his decrepit desk. From this platform he holds that "the history of farming in the Valley is a lack of water and a lack of human replenishment to the agricultural system."[31] Although some Laredoans are proud of their rootedness in the marginal agricultural traditions of the region, only the large or "successful" landowners have incentives to maintain the patterns of poverty and human exploitation that are linked to colonial traditions of place. Hightower has argued similarly that federal agriculture policy, infected by the doctrines of Social Darwinism, reinforces colonial hegemonies.[32]

What locals don't say about the history of their space is, of course, as important as what they do say. With regard to traditional patterns of land use, it is significant that no locals mentioned, or seemed aware, that Hispanic landholding models in the American Southwest have not always been of the colonial hierarchical model experienced in la Frontera Chica. As Marxists are quick to point out, the mode of production employed in la Frontera Chica has changed radically over time. In *Enchantment and Exploitation,* William de Buys documents the prior existence (and recent disappearance) of communal land-holding practices among Hispanic communities in New Mexico.[33] At the Las Trampas Land Grant in the Sangre de Cristo Mountains, for example, farm-land with river frontage was, as in Laredo, granted by Spanish authorities to families in roughly equal amounts. These plots were considered property that could be bought, sold, or inherited. The balance of land within the land grant, however, was known as *ejido* and held in common by the community as a whole. The *ejido* commons was understood to be a community resource to be democratically managed for the common good. *Ejido* lands could not be bought or sold. Even so-called private property in Las Trampas was not as the Anglo communities of North America have constructed it. Private agricultural lands in Las Trampas were not fenced until well after 1917. This land-use prac-tice was a conscious strategy of traditional Hispanic communities to distrib-ute scarce resources fairly within their membership. Families who held little land might prosper beyond their limited means by holding more livestock that would graze upon the land of neighbors. Thus even private property was understood within a community context.

That such a Hispanic tradition existed elsewhere in the Southwest, but was apparently unknown to those who participated in the project, was a missed opportunity. I am not suggesting that the institution of *ejido,* or com-munal landholding practices at Blueprint Farm, would have cured all ills. I am suggesting, however, that the builders of Blueprint Farm might have benefited from knowledge of such precedent. The significance of this lost opportunity will be further considered in Chapter 6, "Reception."

Aggressive laborers, and those who have lost their small family farms, have responded to a history of economic exploitation by migrating season-ally to more attractive labor markets to the north. Locally depressed wages and a high cost of living along the border have conspired to institutionalize migrant laboring as a way of life for those too unskilled to do otherwise. As

Figure 3.14. The Rio Grande separating Laredo and Nuevo Laredo. Author's photograph.

laborers responded creatively to increasing incentives to find better conditions in which to work, landowners responded pragmatically to the evapo-
rating labor market by transforming their capital into more flexible forms.
Some elites simply sold off prime agricultural land along the river for development at prices inflated by the growing trade economy. The Cartesian grid
of the city thus began its intrusion into the agricultural geography originally
responsive to the curvilinear form of the river.

In response to escalating labor demands, many landowners chose to
subvert discontent through other forms of modernization. They did so by in-
vesting in technology and mechanized production. By eliminating some agricultural jobs and professionalizing others, landowners could increase production and consolidate the market. Agricultural economist Willard Cochrane
points out that such consolidation quickly results in the ever-faster pace of
the "technological treadmill," in which farmers either keep up or get out.[34]

It would be incorrect, however, to assume that class boundaries in Laredo
are congruent with cultural filiation. Unlike those of most other Mexican

American border towns, the elites of Laredo—those who dominate both land and business—are of Hispanic origin. Anglos have, however, been integrated into local culture through marriage and business since the late nineteenth century. The result has been a hybrid culture that is certainly not Anglo American, but neither is it Mexican.[35] The hybrid sense of place that pervades Laredo is unique. Where El Paso, for example, is dominated by an Anglo society, Ciudad Juárez, its sister city across the river, is dominated by a Hispanic society. There is no ambiguity about which side of the river one is on. In contrast, "Los Dos Laredos" (as locals refer to these sister cities) are plural. The two downtowns, Laredo and Nuevo Laredo, face each other across the river as a "split center."[36] Local Hispanics experience a sense of cultural continuity from one side of the river to the other that is found in only a few Mexican American border towns like Rio Grande City, Matamoros, and, perhaps, Brownsville. These localized and hybrid qualities of culture in la Frontera Chica are described in the political literature of the border region as *mestiza*—a neither/nor condition that is liberative because it is not subject to the categories of interpretation alternately dominant in the United States or Mexico. In other words, citizens of border regions have greater flexibility to choose and invent practices. Gloria Anzaldúa, for example, claims that *"Los intersticios,* the space between different worlds,"is liberative space.[37] She might argue that those who constructed Blueprint Farm were too constricted by conventional categories and thus failed to dwell in those interstices already in place. The concept of *mestiza* will be revisited in Chapter 6, "Reception."

Laredo is not, however, a neatly bicultural society. Lebanese, East Indian, and Taiwanese families have most recently immigrated to the border region for purely economic reasons. This recent history reflects the continuity of economic conditions that attracted European immigrants, including Jews, to Laredo in the nineteenth century.

The rate of immigration into la Frontera Chica was dramatically enhanced in 1881 by the arrival of the railroad. Its arrival transformed Laredo from being a local agricultural economy to a rail-based transportation center. The

Figure 3.15. The railroad yard at Laredo. In this view looking south toward the international bridge, the town is to the left and the campus of Laredo Junior College is to the right. Author's photograph.

population jump at the end of the century was, until the era of NAFTA (the North American Free Trade Agreement), the largest in the city's history and instantly catapulted the economy of Laredo ahead of rival towns in the region. When a spur of the Pan American Highway also came through Laredo in the 1930's, it cemented the economic role and cultural identity of Laredo as an international transportation center. Trucks, rather than boats or plows, have become the dominant object of local myth and aspiration. As the farm activist Alvaro Lacayo puts it, "Now that we are an international trade center, nobody wants to go back to being a sleepy agricultural town." Today, there is more land-based trade coming through Laredo, in both directions, than at any other town along U.S. borders. Lacayo contends that "The Mayor seems to be saying, 'We'll stop everyone coming through town and by hook, crook, or taxation take a piece of the action.'"[38] There is a consensus among locals that, as the banker Celia Juárez summarizes, the "future of Laredo is in international trade."[39] Some, of course, understand the danger of dependency upon international trade and the Mexican economy. When the dramatic devalua-

Figure 3.16. Tractor-trailers awaiting customs clearance at the international bridge. Author's photograph.

tion of the peso hit the Mexican economy in January 1995, echoing an earlier devaluation in 1981, economic "growth in Laredo stopped," according to Lacayo. "All of Laredo's eggs are in one basket," he claims, "if Mexico gets a cold we sneeze. If Mexico gets an upset stomach, we fart."[40]

Traditional forms of scenographic tourism are not a major industry in Los Dos Laredos, unlike many other border towns. A form of consumer tourism does, however, exist on the U.S. side. The historically protectionist economic policies of the Mexican government created a demand for consumer goods in Mexico that, until NAFTA, went unsatisfied. Conversely, consumer tourism from the American side is stimulated by relaxed policies toward social regulation in Mexico. The historically rigid social policies of the American government—prohibition, for example—created a regional demand for entertainment that remains unsatisfied. The result is the institutional bifurcation of the town. For example, it has become traditional for wedding *ceremonies* to take place on the American side, while wedding *receptions* take place on the Mexican side. Each social practice is relegated to the appropriate place.

Figure 3.17. A site of consumer tourism in Laredo. Author's photograph.

The 1994 invention of NAFTA has, however, amplified questions about the economic function of a border town that emerged with the cross-border *maquiladora* industries of the 1960's.[41] Now that international trade policy no longer requires the physical transfer and storage of commodities crossing the border, it is unclear how the local economy will skim off a "piece of the action." Should the economic frontier disappear, it is unclear to locals how the culture of la Frontera Chica will respond. It is the very uncertainty of the situation, however, that stimulates the expectation of new opportunities and fortunes yet to be made. The citizens of Los Dos Laredos continue to see their fortune as emerging, not from the ground itself, nor from the life-enhancing presence of the river, but from the future economic contingencies of the frontier. The implication of this worldview for the production of place, is, as we shall see, significant.

The architect Rafael Longoria characterizes the border economy from a historical perspective that recalls Agnew's definition of location:

> The efforts to bring about a trade agreement [NAFTA] can be traced to the success of the *maquiladora* industry. The word *maquiladora* was adapted from the Spanish term used to describe the milling of someone else's wheat for a portion of the resulting flour. The current concept is similar to that ancient practice: foreign components are brought into Mexico to be assembled, taking advantage of the abundant supply of inexpensive labor. The finished product is shipped back into the United States, where, upon its reentry, customs duties are levied on the value added abroad.... In effect, *maquiladoras* are a way for Mexico to export its labor without exporting its workers.[42]

The contingent condition of the economy in la Frontera Chica is magnified by the contingent condition of natural resources. Although the landscape of Tamaulipas has been softened by the importation of palm trees and buffle grass, and familiarized by generations of European settlement, it is an unbountiful place—it leaves recent immigrants like Rafael Bernadini "gasping." The Rio Grande, by the time it reaches Laredo, is one of the world's most polluted rivers. Life in such a physical environment is difficult to sustain, culturally and biologically. As in much of the United States, the values of Laredoans reflect, as Vera Sassoon put it, the "throw-away society" in which they live. Daily practices of nature preservation are not exactly commonplace in the lives of locals. Immigrants from Southern Mexico or Southeast Asia have not come to Laredo to preserve this austere natural world, but to exploit a cultural situation that incidentally includes the land itself. William de Buys argues that "the people of pioneer and subsistence cultures everywhere, have constantly underestimated their capacity for injuring the land."[43] Moderns romantically imagine that the rural peoples of any historical period live in closer harmony with the land than do those who live in metropolitan centers. There is, however, much evidence against such a generalization. The subsistence farmers of eighteenth-century Laredo, like their contemporary counterparts, are no exception.

Nor have the old ranching families of la Frontera Chica identified their interests with the natural world that surrounds them. In the eyes of locals like Vera Sassoon, Laredo remains ten to fifteen years behind these times of increasing environmental awareness. Unlike the citizens of other Southwestern cities—Austin and Phoenix come to mind—that depend upon a delicate ecology, Laredoans are, she claims, generally unconcerned with ecological issues.[44] Perhaps ironically, NAFTA has served as an external catalyst to stimulate local concern for degraded ecological conditions. In direct response to pressure from the U.S. government, Nuevo Laredo has improved the capacity of sewage treatment facilities to better match effluent load. Other Mexican cities still dump raw sewage directly into the river because they lack the authority to tax for infrastructure improvements at a level that reflects actual population growth. As a result, most Laredoans still have a decisively negative view of the Rio Grande. There is virtually no public access to the river because it is perceived as a dangerous place—the hangout of drug smugglers, pollution, and disease. Tom Vaughan, a founder of the Rio Grande International Study Center, which now occupies the site of Blueprint Farm, holds that few people on either side of the river are even aware that the river is their only source of drinking water.

Sissy Farenthold, a Democratic colleague of Jim Hightower's and the first woman ever to be nominated as a vice presidential candidate, sees the politics that link the *maquiladoras* and the environment in a particularly sobering light:

In its 1991 study "Border Trouble: River in Peril," the National Toxics Campaign Fund reported on canals originating at U.S.-owned *maquiladoras* in Matamoros that are filled with toxins that pose a daily threat through ingestion, absorption, or contact with the polluted water, or through respiration of deadly chemicals as they evaporate. In Matamoros and other border cities, raw sewage flows in open ditches. In 1991, the American Medical Association declared the border region "a virtual cesspool and breeding ground for infectious diseases."

These problems are not limited to the disposal of human and toxic wastes into industrial canals and thence into the Rio Grande. The problem extends to the transportation of hazardous waste into, and sometimes from, Mexico. Disturbing and even more immediate is the effect of pesticides and contaminated water on the growing U.S.-bound agricultural goods. In a process labeled the "circle of poison" by Senator Patrick Leahy, these products are tainted with U.S.-made pesticides, long since banned in the United States, that are now sent back to the U.S. to be sold.

The 1983 La Paz Agreement, signed by President Reagan and President de la Madrid, was to initiate a new period of respect for and attention to environmental border issues. To date, this agreement has been ignored. It does not have the force of law, and the State of Texas has been particularly disrespectful of its terms.[45]

The increased public consciousness of such pollution of the watershed, and the extended drought of the 1990's, have forced people to acknowledge the need for water conservation measures. The local water master, whose job it is to enforce international water agreements related to the Falcon Dam, is now required to regularly monitor the meter at the intake at each irrigation system to prevent illegal takings. For those on the American side, the daily rinse of the family car and the watering of lawns have become a source of political debate, if not action. Across the river in Mexico, however, the rediscovered water shortage has stimulated a more severe crisis. The government of Tamaulipas has formally asked to renegotiate water rights to the Rio Grande with the U.S. government. Although Tamaulipas is not a desert on either side of the river, modern technology has failed to shield its residents from the desertlike conditions of scarcity. It is not a place to which visitors come expecting to find a health spa.

It is into this local history of space that the personalities and artifacts of

Blueprint Farm were thrown by political circumstance. Their arrival was noticed only by those with a vested interest. When the Texas-Israel Exchange (TIE) was given authority to proceed by the Texas Legislature, Jim Hightower's political needs demanded the selection of a site that would yield instant political income. Members of the Laredo Jewish community were enlisted to locate the site, which turned out to be the old Slaughter farm. This undistinguished but easily accessible piece of property fit Hightower's ideological profile of a parcel of land no longer economically competitive because it was too small to justify the capital-intensive methods of agribusiness. Not only did this property have a history of produce production, but also the owners were in tax trouble and thus in jeopardy of losing their water rights. Out of these politically conducive conditions, a deal was hastily struck to rent the house, and the owners agreed to donate use of the land for experimental crop production. Within four months of occupancy in 1987, the first Israeli farm manager had produced a crop.

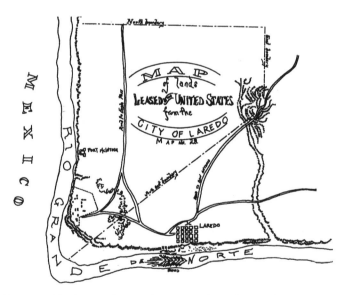

Figure 3.18. Map locating the site of Fort McIntosh at a bend in the Rio Grande. Redrawn from *Map of the Lands Leased by the United States from the City of Laredo,* National Archives, Washington, D.C.

The promise of the jump-started relationship between the Slaughter family and TDA began to exhibit strain, however, as soon as a major grant from the Meadows Foundation made it possible to consider a capital investment in the rented property. When Pliny Fisk began to design the technological systems to be employed at the farm, he assumed that these would be constructed at the Slaughter property. As Alvaro Lacayo characterized the situation, "In lieu of a long-term formal agreement with the Slaughters … the relationship ended in a fiasco."[46] The quickly negotiated lease agreement proved to be a small obstacle to the savvy Mr. Slaughter. Not only did the hastily conceived project have to look for a new home, but the initial investment in the Israeli drip irrigation system was lost in the process—that equipment stayed on the Slaughter farm, where it has remained in productive use.

Almost as quickly as the first site was selected, the project was moved to the campus of Laredo Junior College. The college occupies 200 acres on the river that was once the site of Fort McIntosh, the late-nineteenth-century U.S.

Figure 3.19. Interstate Highway 35 heading south toward the international bridge. Author's photograph.

military post that enforced American sovereignty over the disputed territory that lies between the Rio Grande and the Nueces River to the north.[47] Because the college had no long-range plan, Blueprint Farm was easily accommodated—some have said "dumped"—there. The site of Fort McIntosh and Laredo Junior College is thoroughly isolated from the town by a bend in the river and the railroad tracks that initially brought prosperity to Laredo. A few have concluded that this site was perfect for experimentation because of its proximity to the river. Pliny Fisk also thought it was an ideal site because, on state-owned property, the experimental technologies proposed would be exempt from local building code compliance. As we shall see in Chapter 7, however, such pragmatic criteria also exempted the "demonstration" from a principal responsibility—the ability to be witnessed.

NARROWING
HORIZONS OF
SPATIAL DISCOURSE

The history of space produced by the inhabitants of la Frontera Chica can now be interpreted through those lenses donned in the first section of this chapter. As in Lefebvre's discourse on the production of space, I want to now argue that the history of local space "begins . . . with the spatio-temporal rhythms of nature transformed by social practice."[48] Lefebvre's emphasis upon "practices," rather than objects, supports the organization of Figure 3.4. If the reader will recall, in that figure I position "human practices" as central to the social production of both places and technologies. Based upon the structure of that figure, the conclusion of this chapter will consider the technological and spatial practices of Laredoans in relation to John Agnew's characterizations of place: *location, locale,* and *sense of place.*

The historical *location* of Laredo—Agnew's first characteristic of place—begins with the Spanish colonial Orders for Discovery and Settlement of 1573. It is essential to recognize that this ordinance was conceived as an instrument of sixteenth-century global political economy. In Lefebvre's analysis of that ordinance, he argues that

The very building of the [Spanish colonial] towns thus embodied a plan which would determine the mode of occupation of the territory and define how it was to be organized under the administrative and political authority of urban power.[49]

In Lefebvre's terms, the Spanish ordinance is best understood as an instrument of production—a technological code that served as a means to impose an emerging proto-capitalist order upon a premodern space. It is in this sense that the colonial ordinance was a "representation of space"—what Lefebvre refers to as a kind of mental or conceptual map that directed the emplacement of the emerging European mode of production on top of that practiced by indigenous Americans.

A geography of technology concerned with location, or the structural conditions of political economy, is naturally concerned with competing territorial claims of European (and American) colonial powers. The mapping and remapping of Laredo by the four nations that have claimed the city exemplify these concerns.[50] Such political instability documents the stresses of political economy to which the region has been historically subject. I have been tempted to periodize the location of Laredo in the global economy as pre-1881 and post-1881, that being the year when the railroad reached the city. The early period—before 1881—was generally dominated by the successive colonial economies to the south, and the later period—after 1881—has been generally dominated by the market economy to the north. Such a rigid periodization, however, would tend to obscure the continuum of spatial development in the region. It would be better to argue that la Frontera Chica has been a transitional place in the evolving modern political economy— alternately claimed by competing national interests. Its current status as the "Gateway to Mexico"—the largest land-based port of entry into the United States—is thus not a newly minted condition.[51] My argument here is that *the location of Laredo has always been a function of the globalized economy.* What is changed is not the local presence of economic powers acting at a distance (what Latour describes as "centers of calculation"), but the relative location or

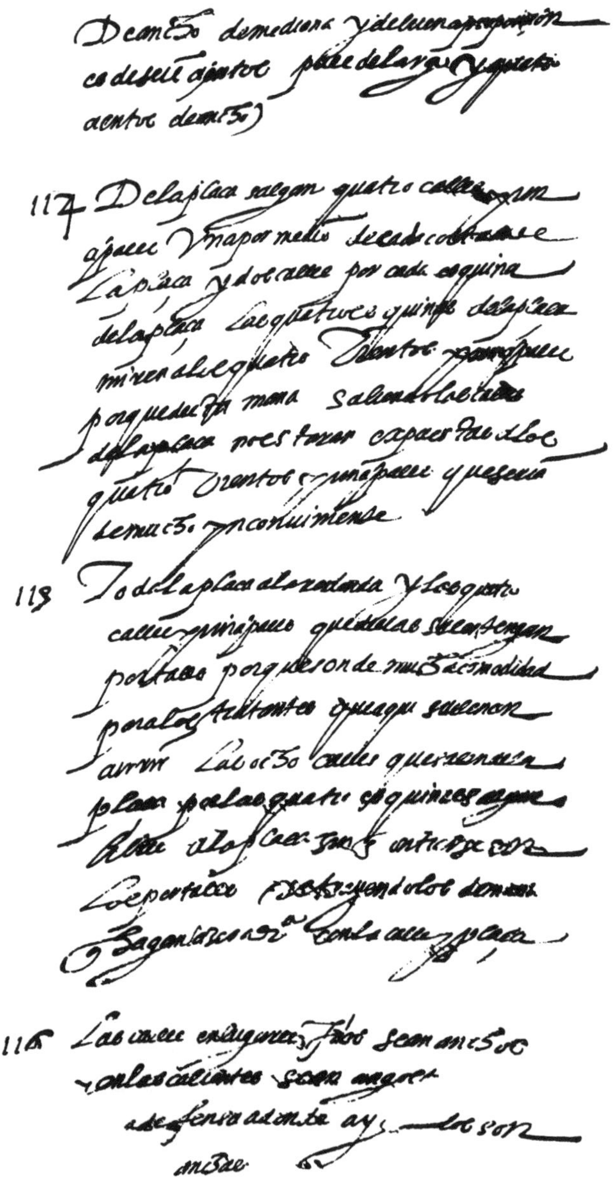

Figure 3.20. Directions for the making of a town square in the New World. From Ordenanzas de descubrimiento, nueva población y pacificación de las indias dadas por Felipe II, el 13 de julio de 1573, Archivo de los Indios Sevilla. The English translation is generally cited as Spanish Orders for Discovery and Settlement of 1573, the original of which is located at the Archives of the Indies at Seville, Spain.

strength of those powers. Rather than being tied to the court of Spanish monarchs, the French Empire, or Mexican revolutionaries, Laredo is now linked most strongly to the stock exchanges of New York, Frankfurt, and Tokyo.

A geography of technology concerned with *locale,* Agnew's second characteristic of place, is one concerned with the "setting for social relations." In the case of Laredo, the original eighty-nine Spanish land grants, or *porciones,* granted to settlers by Spain in 1767, created what might be described as a binary landscape. Each *porción* surveyed by colonial authorities included two parcels: a residential plot within the colonial grid of the town and a deeply linear, agricultural plot with narrow river frontage that included water rights.[52] The patterns of property rights and the daily practices of citizens were thus ordered by a distinct separation between two kinds of space: the warped linear space of agriculture and the checkered space of society and trade.

As John Stilgoe has documented, the premodern European *Landschaft* also observed a distinction between agricultural and social space in that villages were densely clustered settlements surrounded by agricultural lands.[53] In that landscape, however, both agricultural and social spaces radiated organically from a central locus experienced by the community as a common spiritual center. In the colonial Spanish ordinance, however, spaces allocated for agriculture and society were conceived quite differently. Although social space was still surrounded by agricultural space, the mode of ordering these practices was geometrically and geographically distinct. The social space of the village was centered upon the abstract space of political, rather than spiritual, authority, and agricultural space was mathematically surveyed in relation to whatever local resources determined economic potential. In other words, agriculture in New Spain had already become conceptually, and thus spatially, disengaged from those more abstract political and economic practices assumed to dominate public life. Where premodern European space was perceived and lived as a *monist,* or singular, geography, the modern space of the New World was conceived as *binary,* or Cartesian.

In the period after the coming of the railroad to Laredo in 1881, interrelated networks of railways and highways have increasingly dominated the conditions of daily life and commerce. Those agricultural practices that once engaged the river to produce use-value have gradually been replaced by commercial practices that move people across the river to produce exchange-value. Laredoans—other than the elites who own the land—have generally rejected the stasis of farming and rootedness along the river as a way of life. What Laredoans say they want—at least the poor Hispanic families who eke out a living by working in the *maquiladoras,* tenant farming, or migrant laboring—is to escape the dark social history of their bondage to the land. As a locale, Laredo embodies the tension between stasis and movement, between farming and flight. The periodization of Laredo as pre- and post-1881 may be best portrayed as identifying a watershed in a continuum of change. It is reasonable to hold that in the early period the river was the conceptual center of the community. In that period the river was the net that bound the community to the natural cycles of agricultural production. In the later period, however, the river has become both a conceptual and real boundary—a marker that delineates changed social and economic conditions. The network of rails and roads, in the period after 1881, rather than centering the community upon cycles of agricultural production, has centered the community instead upon unequal rates of exchange that are determined by others at a great distance.

In the very real sense illustrated by Figure 3.21, the ever-advancing commercial and suburban residential development that emanates from the Cartesian geography of the colonial city has gradually subsumed the agricultural geography of the river. The original binary geography that characterized the setting for social relations in the town has been transformed over time to a new monist physical geography—one based upon trading rather than agriculture as a way of life. My point is that the binary space established by the Spanish colonial ordinance in the sixteenth century was proto-modern. I mean by this term that the eighty-nine *porciones* surveyed in 1767 were a conceptual departure from the premodern pastoral monism of European space, but were not yet identical to modern capitalist monism. The space surveyed in

Figure 3.21. Contemporary aerial photograph of Laredo. Nuevo Laredo is at the top of the image on the south bank of the Rio Grande. The site of Blueprint Farm is at the extreme right, or west across the railroad track from the city proper. Note how the pattern of land use changes to the west of the tracks. The Cartesian land-use pattern along the river that is derived from the colonial *porciones* initially allocated for agricultural purposes in 1767 (as illustrated in Figure 3.9) has seamlessly accommodated commercial and residential development. Photograph by Danny Alcocer © 1996.

the New Spain of 1767 was somewhere in between these singular visions of place. By monistic space I do not mean homogeneous space. The premodern space of the European *Landschaft,* like modern urban space in twentieth-century North America, was distinct and varied. By the term monism I mean only that the production of space was driven by a dominant idea, or set of concepts. In the case of Laredo, it was the Spanish who imported the seed of capitalistic monism that now dominates space on both sides of the Rio Grande.

I want to stress that the shift to a modern version of spatial monism in Laredo was not inevitable. Like the Spanish *pobladores* of the Las Trampas Land Grant in New Mexico, the inhabitants of la Frontera Chica might have made different technological choices. In New Mexico the colonial grid of densely settled urban, Cartesian space was generally abandoned in the early nineteenth century as soon as the threat of Indian attack had abated. Abandoning the colonial grid allowed these pastoralists to scatter their houses "along the edges of irrigation fields so that each family might better guard its crops from livestock and theft."[54] An organic, noncapitalist monistic space has been the result. Although the settlers of la Frontera Chica and Las Trampas began with the same binary geography as a conceptual map from which to build, they have over time produced very different geographies indeed. Where Laredo has become relentlessly competitive and capitalist, Las Trampas is characterized by *vergüenza*—a sense of "self-effacing probity" that restrains members of the community from advancing their own interests over those of the community.[55] Although the conditions of nature and economic forces

Figure 3.22. A liquor store on the Mexican side. Author's photograph.

acting at a distance certainly limited the choices made, each community might have made other choices. The Laws of the Indies did not predetermine the practices that sprung up in those places.

Finally, a geography of technology concerned with the *sense of place*, Agnew's third characteristic of place, is concerned with the "structure of feeling" of that place. Before its international partition in 1846, Laredo was a centered pastoral town on the north bank of the river. After partition, those who wished to retain their Mexican citizenship settled in Nuevo Laredo on the south bank, which was soon thereafter chartered as an autonomous Mexican municipality. Although Laredo is unique among U.S.-Mexican border towns in that both towns share a political structure and cultural identity dominated by Hispanics, the unequal economic forces at work in Los Dos Laredos create the sense of a split center. In this sense, the international city is still a binary construct, but one very different than the divided land-use patterns imported by the Spanish. On the north bank of the river, commodities are acquired and exchanged; one can almost hear the meters ticking. On the south bank of the river, an entertainment market has been created by the disparity between Mexican and American attitudes toward social regulation. Here the meters have a mariachi tempo. Where the colonial bifurcation of Laredo's locale structured how one moved *through* space, the more recent bifurcation of Laredo's sense of place structures how one thinks and feels *in* space—it is like the right and left side of one's brain.

Lefebvre might summarize this chapter by arguing that, in the "long history of space," the forces of political economy increasingly abstract the anthropological origins of daily life.[56] The trajectory of history, in Laredo and elsewhere, has generally been from the "absolute space" lived by archaic peoples to the "abstract space" surveyed by moderns. Because Laredo is a very old city, at least by the standards of the American West, we moderns tend to romanticize it as a place somehow different, or other, than those suburban landscapes now constructed by flexible capital. In Lefebvre's terms, however, Laredo was never a romantic assembly of "absolute spaces" linked to the spiritual practices of premodern life. Nor was it ever a mission intended to pacify souls and plow the earth so much as it has been a geographic opportunity created by displaced humans crossing the river to produce exchange value. In this sense, the colonial grid imposed upon Laredo in 1767 can be best understood as the technological apparatus through which the premodern space of Laredo was re-*located* by the Spanish masters.

If "every space has a history, one invariably grounded in nature," as Lefebvre argues, then the ecological conditions of Laredo, and the cultural conditions of la Frontera Chica, have not prefigured an agricultural future.[57] Nor have they precluded such choices. The ecological conditions and history of technological choices made in the Las Trampas Land Grant of New Mexico have been quite different. Similarly, where other towns in la Frontera Chica, like Roma, have become increasingly detached from centers of calculation through the weakness of their technological links, or through the strength of those who have made difficult technological choices, Laredo has reinforced its colonial potential as a trading town. It has done so by being better positioned to receive stronger technological links—the railroad, bridges, and Pan American Highway are only the most obvious examples. This argument does not erase the real, but marginal, agricultural history of la Frontera Chica. Rather, it acknowledges the principal point of this chapter: that most people have immigrated to Laredo, pre- and post-1881, to take their chances at trading— in the broadest meaning of that word—as a way of life. The space of Laredo has been increasingly abstracted by market forces acting at a distance and by those migrants who are passing through on their way to prosperity.

This argument has, as we will see, significant implications for the construction of an experimental farm. I want to be careful, however, not to leave the reader with the sense that Laredo has been transformed into an undifferentiated space indistinguishable from, say, suburban Columbus, Ohio. This is obviously not the case. If the location of Laredo is increasingly abstract, locals with deep roots in the city increasingly articulate its sense of place as "mestiza"—meaning that it is a place resistant to appropriation by the mechanisms

Figure 3.23. Blueprint Farm as it appeared in 1995. Author's photograph.

of flexible capital. This unique sense of place, I will later argue, is a lost opportunity for the construction of an experimental farm.

In sum, it is worth repeating that Spanish colonials conceived Laredo as a binary geography that spatially alienated the practices of agriculture and trade. In the two centuries or so that have followed the deployment of that conceptual diagram, the binary quality of space has been gradually modified by the technological choices made by those immigrants who have been attracted to the political economy of the borderland. In the long public discourse between those who would farm and those who would trade for a living, the voices of traders have been alternately louder and better supported by distant networks. As a result, they have dominated the long history of local space in la Frontera Chica. In other locales, such as Las Trampas in New Mexico, farmers have prevailed, if not prospered. What those differentiated spaces share, however, is the dialogic mode in which technological choices shaped space.

To conclude this chapter I want to generalize the dialogic relationship of technologies and places that emerges from this particular story. The first step is to state that the qualities of place and technology are not interchangeable. They characterize different things, but the *location, locale,* and *sense of place* that describe Laredo as a place are largely congruent with the *human knowledge, human practices,* and *sets of physical objects* that describe the competing technologies that operate there. In the end, it is difficult to distinguish the boundaries of technologies and those of places. On the ground, these concepts are less distinct than assumed by the conventions of modern thought. This argument leads me to a general hypothesis:

Places and technologies are different things, but the processes of their social construction are dialogically related.

To clarify the distinction that I want to make between "dialectic" and "dialogic" relations it will be helpful to offer the reader two formal definitions. The definition of the dialectic is taken from J. K. Gibson-Graham's citation of Louis Althusser, who was in turn citing Marx:

[the dialectic] includes in the positive comprehension of the existing state of things at the same time also the comprehension of the negation of that state, of its inevitable breaking up; because it regards every developed form as in fluid movement and thus takes into account its transient nature, lets nothing impose upon it, and is in its essence critical and revolutionary.[58]

My operating definition of the "dialogic," so as to amplify similarities and differences to the Marxian dialectic, is simply a modification of the above:

> The dialogic includes the positive comprehension of the contested meaning of things, and at the same time, it also includes the will to move that state toward a cultural horizon of meaning. Because the dialogic regards every developed form as a fluid movement, and thus takes into account its transient nature, it invites all to contribute to its development, and is in its essence life-enhancing and revolutionary.

This definition, cumbersome as it is, will help the reader to interpret my use of the term as we investigate the topics of *intention, technological intervention, reception,* and *reproduction* in Chapters 4 through 7. Chapter 8 will more clearly set out the practice-based propositions that derive from the dialogic hypothesis stated above.

CHAPTER FOUR

CONFLICTING INTENTIONS

...no one lives in a culture, shares a paradigm, or belongs to a society *before* he or she clashes with others. The emergence of these words is one consequence of building longer networks and of crossing other people's paths.

Bruno Latour, *Science in Action,* p. 201

This chapter investigates the converging and diverging intentions of those who produced Blueprint Farm. In the first section I investigate the concept of *intentionality* as a formal, or philosophical, tradition. This section provides a background for better understanding the competing intentions of those who produced the farm. These competing networks of intention are reconstructed in the second section. In the third section, "Uninhabited Intentions," I reconcile the empirical evidence gathered from the farm with the formal concept of intentionality. This logic concludes in a novel definition of technological determinism that will prove helpful in Chapter 5, where I investigate the *technological interventions* of those who produced the farm.

In conventional conversation, to "intend" is "to have in mind: to plan," or "to design for specific purpose or destine to a particular use." This conventional meaning seems quite clear. We make plans for things, be they objects, events, or practices. The etymology of the word "intend," however, derives not from the concept of *planning*, but from the more elusive concept of *stretching*.[1] In archaic use, to "intend" must have implied the stretching of one's mind into the material world beyond the skull. Such a history of meaning suggests that contemporary intentions capture an ancient tension between that which is inside our minds and that which is outside our bodies. In the world constructed by moderns, intending links the competing interests of human subjects to the ever more scarce assets of nonhuman objects.

To recognize that social projects often enlist those with conflicting intentions is hardly an original observation. The plans that we have "in mind" are rarely congruent with the minds and plans of others. At least at first. In normative practice, planning is commonly understood as bringing the conflicting intentions of *producers*—meaning architects, engineers, clients, government, lenders, and others—"into line" with their collaborators so as to mount a unified campaign to build something. Architects, of course, play a central role in narrowing the conflicting intentions that link producers to a common project. The question remains, however, if the horizon of intention negotiated by the architect is focused upon a *human practice* or an *object*. In the case of Blueprint Farm, the evidence suggests that a few producers attempted to satisfy their intentions by living differently—they intended to construct a setting for human practices. Most of the farm's producers, however, attempted to satisfy their intentions by literally building them in material form—they intended an object. In this chapter I'll argue that it was, in part, the confusion between practices and objects, or between activities and things, that led to tragic conclusions.

The conventional use of the term *intention*, noted above, suggests a complex history. That history of meaning has been investigated by a series of thinkers (including Franz Brentano, Edmund Husserl, and Martin Heidegger) who have argued that all

THE CONCEPT OF INTENTIONALITY

human activity is characterized by a kind of directedness, or intention, toward objects.

It was Brentano who first articulated the theory of intentionality. Although he returned to this topic again and again over the course of his career, he never succeeded in adequately describing the elusive nature of the *intentional object* as distinct from the common physical object. Brentano searched for a way to get beyond the Cartesian distinction between the *idea of the thing* and *the thing itself,* but was unable to satisfactorily define the phenomenon of directedness he claimed to experience.

It was Husserl who formalized the distinction that so eluded Brentano. Husserl's move was to synthesize the Cartesian duality of subject/object into a third category, the *noema,* that is said to exist between the mental and material states. In Husserl's construct, the noema accounted for the directedness of mental phenomena toward the material. Husserl argued that "the directedness of the act should not be accounted for by some object toward which the act is directed, but a *certain structure of our consciousness* when we are performing the act" [emphasis mine]. The noema constructs consciousness "as if" there were an object, but does not require one.[2] For Husserl, consciousness can no longer be considered a "self-sufficient and self-contained domain of interiority," as is understood in Cartesian philosophy, because consciousness itself is directed toward, and thus dependent upon, exterior objects. Aron Gurwitsche interprets Husserl's version of intentionality in such a way that the noema is not confused with the Cartesian concept of "idea"—a representation of, or substitute for, reality. Rather, noema is the *structure of consciousness* itself that is directed toward a satisfaction inscribed in the object. Noema is not the representation of a physical thing.[3] Rather, "intentional states have as their content a representation of their conditions of satisfaction.[4] Nor should an "intentional state" be confused with a mental state. The noema is defined by Gurwitsche, not as a psychological event, but as that which should be understood to be "the object as it is intended."[5]

Husserl's reconstruction of the duality of interior and exterior (that which we have "in mind" vs. that which we make plans for), or of subject and object, constitutes a major critique of the doctrines of Cartesian philosophy. For Martin Heidegger, however, Husserl's construction of intentionality, like Brentano's before him, remains within the Cartesian subject/object dualism. Heidegger's critique of the phenomenologists is that they still depend upon a transcendental mental act to transport the knowing human subject out of her interior isolation and into the world of things. In Hubert Dreyfus's interpretation, "Heidegger accepts intentional directedness as essential to human activity, but he denies that intentionality is mental, that it is, as Husserl (following Brentano) claimed, the distinguishing characteristic of *mental states*."[6] Dreyfus, then, rejects Gurwitsche's argument about the distinction that might be made about the difference between "intentional states" and "mental states." Rather than depend upon some mediating "structure of consciousness" to overcome mentalism, as do Husserl and Gurwitsche, Heidegger and Dreyfus construct a radical ontology that replaces epistemological questions altogether.

In Husserl's version of intentionality, "meaning-giving *knowing* subjects" construct a relation to objects through a reflective act of "phenomenological reduction." In Heidegger's version of intentionality, "meaning-giving *doing* subjects" are always already engaged with a network of social practices that engage other humans and nonhumans.[7] For Husserl, subject/object relations are constructed. For Heidegger, subject/object relations already exist within the historical continuum of *background practices* in which we find ourselves. It is these background practices that constitute for the individual in society a *preontological understanding* of Being in a place. In Heidegger's view, *knowing how* things relate to each other ontologically is more important than *knowing that* they relate to each other epistemologically in a certain way. The radical ontology that Heidegger articulates in *Being and Time* thus provides a workable alternative to the version of intentionality constructed by Brentano and Husserl, and to the Cartesian dualism that shaped the modern understanding of subject/object relations.

Dreyfus offers an analysis of "distance-standing practices" in different cultures as a concrete example of how preontological understanding shapes the relationship of human bodies in space. He observes that Finns, for example, stand farther away from each other during common conversation than do Americans. Americans, in turn, stand farther away from one another than do Italians. Such background practices are not learned in any conscious way; rather, they accumulate through social experience. They are what some would define as "tacit," or unspoken, knowledge. This simple example serves to illustrate how the social construction of shared practices constitutes the relationship between things in the world. From this example of body relations we can generalize about the relationship between other human and nonhuman entities. The point is that, for Heidegger, it is the *relation between things,* rather than the *subject's consciousness of things,* which characterizes Being in a place. If the reader will recall Lefebvre's contribution to the previous chapter, he argued that "space [or place] is not a thing, but a set of relations between things." We can thus associate Lefebvre's production of space with Heidegger's version of intentionality.

Before concluding this section, I want to distinguish between Heidegger's philosophy of relations, which is articulated in *Being and Time,* and his philosophy of technology, which was produced almost thirty years later. It is the early work that opens a non-Cartesian alternative that Bruno Latour uses as the ground to develop the nonmodern philosophy of relations upon which my own work relies. This nonmodern alternative is distinct from, and opposed to, the postmodern philosophy of relativism that derives from Heidegger's later poetic work. More will be said about these distinctions in Chapter 5. For the moment the distinction provides an opportunity to introduce Latour's concept of the "quasi-object."

If we accept Heidegger's ontology, we are left with a radically altered relationship between humans and nonhumans. In the place of humans as knowing subjects, and nonhumans as available objects, we are left with a field of "quasi-objects" and "quasi-subjects." In different situations, subjects become objects, or conversely, objects become subjects.[8] This is not a situation unique to humans, who are alternately knowing subjects, or, say, the known object of some bureaucratic, medical, or technological treatment. Nonhumans, too, switch from quasi-objecthood to quasi-subjecthood. Ecologists recognize, for example, that nearly all organisms do more than merely adapt to the inevitable conditions of nature. In fact, organisms of all types *create* the "natural" conditions in which they prosper. In this sense animals are active subjects who transform nature according to its laws.

To define the nature of such relationships as that between quasi-objects and quasi-subjects, Heidegger used the term *Verhalten,* or "comportment." This term is used to describe the way in which human activity is directed because it has no "mental overtones." In other words, our comportment toward things, or toward the environment that we shape, does not require consciousness of us. In Heidegger's ontology, we comport ourselves toward the things that inhabit our world in a consistent manner. This purely philosophical proposition thus leads, however indirectly, to a question regarding the case at hand.

How, then, did those who produced Blueprint Farm comport themselves toward their project? How did they comport themselves toward the objects that they created?

I have identified five sets, or networks, of competing intentions that were at work on the farm. These networks cooperated just enough to physically construct Blueprint Farm, but not enough to sustain it. In this section I categorize these five sets of intentions as *technological networks.* Pliny Fisk III frequently uses the term "network" to describe systems of information and energy flow. Bruno Latour's term "technological network" (discussed in Chapter 3) is, however, a richer concept for my purposes here. In describing the nature of technological networks, Latour generally refers to collective "interests" rather than "intentions." For the sake of clarity, I would like to distinguish between these terms by holding that interests are fully conscious, while intentions are not. Our conscious interests, then, are potentially affected by a

**NETWORKS
OF INTENTION**

wide range of social forces, resulting in something like what Marxists refer to as *false consciousness*. My point is the interests expressed by a group might then be different than their intentions.

For Latour, the term technological network describes not just the sets of common interests that bind humans together, but also the unconscious relationships between human interest groups and resources, or the relationship of subjects and objects in a place. Heidegger's philosophy of relations and his version of intentionality thus provide a conceptual foundation that supports Latour's terminology. In this Heideggerian sense, the technological network is an essentially spatial concept that links particular human intentions to the objects found in particular places.

There were five networks that collaborated to construct Blueprint Farm. These were: First, the Hightower network, which included the constituents and supporters of Jim Hightower both within and outside the Texas Department of Agriculture. Second, the Israeli network, which included local Jews as well as Israeli political and agricultural interests. Third, the local expert network, which included local business and education interests. I want to state clearly that this group of local experts, although they presumed paternalistic authority to do so, did not represent the interests, much less intentions, of local farmworkers. Fourth, the land grant network, which included the combined interests of the land grant universities, agricultural equipment producers, corporate growers, corporate consumers of produce, and biotechnology producers. Finally, fifth, the ecologist network, which included the activist architects, labor organizers, and organic gardeners. Each of these networks, including the local expert network, was global in scope. I mean by this that none of the competing interest groups had strictly local, grassroots interests in mind. All had extended interests at stake in the making of Blueprint Farm. To make those explicit will help us to better understand the technologies and place constructed at the farm. In the remainder of this section I'll briefly review the interests of each network. The underlying intentions of each network will be made explicit in the final section of this chapter.

In previous chapters enough has already been said about Jim Hightower's instrumental construction of common cause with Jewish interests. When asked directly about Blueprint Farm and his connection to Israel, Hightower claimed that

> The intent of TIE [the Texas-Israel Exchange] was to see if we could do something productive with Israeli technology because Israel has similar weather and geological conditions as Texas.[9]

This is an ideological affiliation with the doctrines of bioregionalism. In the same discussion, however, Hightower reasoned that "the farm was a very pragmatic expression of TIE's intent—*real Israeli input!*"[10] Hightower's candid emphasis on "Israeli input" is in distinct contrast to the "mutual exchange" of input supported by the Texas Legislature and anticipated by such locals as Alvaro Lacayo. But according to Hightower, his connections to the Jewish community, and to Laredo, were entirely circumstantial. In both cases, he insists that the contact was instigated by Sarah Ehrman, a Department of Agriculture staff member who later became the first director of TIE. Participants in the project, however, understood Hightower's link to the Jewish community to be instrumental, rather than circumstantial. Some understood the relationship in the classical terms of populist politics and the development of a diverse "political power base."[11] Others, however, were well aware of the political shadow under which the connection to Jesse Jackson had placed Hightower. These participants later became bitterly cynical about their leader's commitment to what Lacayo described as "Israel and glory." Ecologists in particular became increasingly suspicious that the decision-making procedures of the Hightower regime existed only to promote Israeli interests. This suspicion, however, should be understood in the context of Hightower's specific policy intentions.

Hightower's stated policy intention was to "stimulate the economic viability of small landowners."[12] In tactical terms, this policy was designed to divert resources from the "good ol' boys"—the large landowners who had been served so well by the land grant network—to small landowners. Hightower's "four-point program" to redirect resources included: (1) the introduction of crop diversity, (2) direct marketing, (3) conservation of resources, and (4) value-added production.

The diversification of crops, the first priority of the new TDA policy, was designed to combat not only the vegetative monoculture propagated by agribusiness, but to stimulate high-profit, ecologically suitable "niche crops" that are uniquely manageable by small landowners. Such "niche farms would not be stupid enough to grow corn" and compete in the global commodities market. Rather, the interests of small farmers lay, Hightower argued, in a diverse ecology and flexible market strategy.

It was not by accident that TIE was located in the Marketing Division of the Department of Agriculture. Even the hostile rank and file of the Agricultural Extension Service (AES) understood that the second point of Hightower's program—direct marketing—was intended to threaten the economic hegemony of the large supermarket chains. H-E-B, the chain that dominates the South Texas market, was a prime target. By developing a system of direct retail markets, through which niche farmers could sidestep their dependence on H-E-B's distribution system, Hightower hoped to economically emancipate the small farmer. Such a limited field of engagement against the land grant network seemed more prudent than entering the far more controversial battleground of property rights and the distribution of productive resources.

Hightower's third policy item, the conservation of resources, developed more slowly. Initially his interests were not only to conserve water, but also to minimize the use of pesticides and herbicides. In the period 1984–1986, when TIE was articulating its program, the concept of *sustainability* was just emerging at TDA. It was, in fact, critics from inside TDA that forced the emergent concept of sustainability into the official rhetoric of the project. In time, the language (if not the practices) of sustainable development replaced the simpler concept of resource conservation. I should stress here that simple resource conservation and sustainability have come to mean different things. When Blue-print Farm was conceived, however, the definition of those terms was unsettled. It had been Hightower's intention to move on ecological issues "one step at a time." By his own admission, "Fisk's concepts of sustainable building design came later."[13] The political opportunity to use the rhetoric of sustainability was, how-ever, an opportunity that Hightower embraced quite early.

The sociologist Aant Elzinga makes the significant distinction between "practical-instrumental research" and "symbolic-instrumental research." The former is aimed at solving immediate, tangible problems. The latter serves ends that are primarily political and may influence nonscientific fields. In the sense understood by Elzinga, "symbolic-instrumental research" is ironic in that its in-tentions are directed at other than the "scientific" product. On the basis of the evidence, it is reasonable to argue here that Hightower's intentions toward Blue-print Farm were "symbolic" and "instrumental" rather than "practical."[14]

Hightower's fourth policy item—value-added economy—required that TIE and TDA conceive ways for small farmers to process agricultural produce as high-end finished goods. Tomatoes, rather than being sold in bulk for sauce, could be sun-dried and marketed at dramatically higher prices. Flowers, rather than being sold in quantity to florists, might be arranged or dried. By all ac-counts, however, such market strategies never materialized, except in the marginal schemes of Fisk and his ecologist collaborators.

As downhome and pragmatic as these four policy tactics may seem, Hightower's program did not emerge from long years on the farm. Jim Hightower entered the agricultural politics of Texas, not through any of the tra-ditional routes connected to the land or industry, but through his tenure in Washington as aide to U.S. Senator Ralph Yarborough. Hightower's visible net-work was more closely linked to the remnants of the LBJ legacy than to back-slapping on the back four hundred. It was this sophisticated distance from local farm practice that made many suspicious of his long-term investment in re-gional farm issues. Some suggest that their suspicion of Hightower's instrumen-tal populism has been vindicated by Hightower's activities since leaving office.

Even Hightower himself acknowledges that he has no continuing involvement with agricultural policy in Texas or elsewhere. Then, as now, his interests were, by local standards, political and abstract. The intentions of Hightower and his supporters were directed not just at small Texas landowners, or at the Jewish community, but at national (or global) ambitions. Many have speculated that he lost the election as Texas Commissioner of Agriculture in 1990 because he was "stretched" too thin by plans to run for the U.S. Senate.

Hightower's global intentions were seen as both an asset and a liability to his local authority. By introducing new ideas to local conditions, he made both friends and enemies. It was, of course, this abstract view of the local and the universal that enabled him to understand the potential ecological and political links that might be constructed between Texas and Israel. Jim Hightower's stated interests were radically democratic. At the local level of political relations, however, his interests were conducted instrumentally. The tragedy is that the instrumental means that Hightower employed to establish the regime of sustainability co-opted his democratic intentions. Hightower's admirable democratic interests were served by architectural and agricultural research that was symbolic and instrumental rather than concrete and ontological.

The Jewish Network   Most Laredoans assumed that the local Jewish community sponsored the Israeli presence in town, not TDA or Jim Hightower. Local Jews themselves saw the Israeli connection as an unexpected opportunity to cement ties to what some described as the "old country." The Israeli farm manager was "wined and dined by local Jews," who in turn furnished the homes of style-conscious matrons with a steady supply of cut flowers.[15] This happy arrangement with Israel was welcomed by the local Jewish community because it was an opportunity to "politic" in the arena of land policy, normally closed to their interests by large landowners.

For local Hispanic activists, like Dr. Héctor Jiménez, the Israeli connection made subversive political sense. Jiménez perceived that Israelis, like Laredoans, "had limited resources."[16] This association put the economic insecurity of Jews and Hispanics at odds with the apparent wealth of the Texas agriculture industry, or what Hightower called the "land grant complex" and which I refer to as the land grant network. (Hightower used "complex" to relate the concept to Dwight Eisenhower's term "military-industrial complex." I use "network" to relate the concept to Latour.)

Of equal significance, limited Israeli resources were directed toward "trying to grow as much as possible on a small plot of land"—a practice that the land grant complex openly worked against by supporting land consolidation and industrial farming. There was considerable irony in the fact that locals perceived the Israelis, who came from halfway around the world, to be "a better bet" to extend Hispanic interests than the land grant network concentrated a hundred miles away at Texas A&M University.[17] Of course, local Hispanic activists also understood that the flip side of Israeli inventiveness was the deep pockets of the Texas Jewish community.

When the Israeli team of agricultural scientists arrived in Laredo to study the feasibility of the project, they imported—along with the drip irrigation technology—an aura of technological superiority. Locals imagined themselves to be the recipients of a "phenomenal and dramatic new technology," as if it were being delivered in a box tied up in garlands.[18] At this early stage, it was clear to Laredoans and Israelis alike that the purpose of the project was to "demonstrate Israeli technology."[19] But most sophisticated locals, like the labor activist Alvaro Lacayo, also understood that the Israelis were not Peace Corps volunteers. Lacayo holds that "They [Israelis] were here to extend their shores, to set up a marketing system for their technology."[20] The Israeli irrigation technology was, in fact, a very sophisticated and packaged commercial enterprise. The donation of an integrated system that included computers, greenhouses, pipes, valves, chemicals, and personnel by Israeli manufacturers was an investment in market expansion, not the manifestation of a desire to meddle in the politics of American agriculture and property rights, or engage locals in philosophical discourse.

Along with others, Alvaro Lacayo came to understand that "What excites Israelis is that they made the desert bloom. That has nothing to do with sustainability."[21] Or at least according to the definition emerging among local ecologists. Over time, the Israeli intentions toward the land became clear. Not only did the drip irrigation system depend upon the massive use of chemical fertilizers, pesticides, and herbicides, but "the Israelis never had any intention of farming organically."[22] Israelis were in Laredo to expand a technological network by winning new and influential supporters. Local Jews saw themselves as the cultural beneficiaries of Israeli technological dominance. So long as cultural goodwill was the output, there was no need to examine the contents of the imported black box. Jewish interests were served by the techniques of commerce.

The Ecologist Network

The ecological activists who collaborated in the "building project" represented three radical constituencies: architecture, organized labor, and the organic gardening movement. As a whole, these collaborators understood their common cause, as articulated by Fisk, to be a program of "social change."[23]

As a designer, Pliny Fisk III has no intention of practicing architecture in any normative, or professional, context. For Fisk, "architecture is activism … a tearing into the status quo" of the profession itself. He sees his work as participating in the radical anarchist traditions of Murray Bookchin, social ecology, and Gaia politics.[24] Rather than inventing a "signature style" of architecture for consumption in the market, Fisk has worked since 1975 to invent the Center for Maximum Potential Building Systems, "a nonprofit education, research, and demonstration organization committed to bridging the public and private sectors and linking policy initiatives with practice."[25] In Texas, nationally, and internationally, Fisk has become something of a guru in the sustainability movement. His network of professional contacts reads like the pages from a *Who's Who in Environmental Politics*.

One of these contacts, extending back to the early 1970's, was Alvaro

Lacayo, a labor organizer from Laredo. Unlike Fisk, Lacayo has a long history of connection to active leftist Texas politics that gave him access to the Hightower network. The string of personal relationships that tied Lacayo to Hightower's interests included ties to the United Farmers and Ranchers Congress, the Federation of Southern Cooperatives, SNCC (the Student Nonviolent Coordinating Committee), and CORE (the Congress for Racial Equality). After Hightower's restaffing of TDA in 1981, Lacayo found, much to his surprise, some of his old friends working for state government in Austin. Such relationships only confirmed Lacayo's commitment to Hightower's program for Texas agriculture. He and Fisk collaborated on plans for farmworker housing on the Blueprint Farm site and a number of other "outreach" projects that might "change how people live" in la Frontera Chica. For Lacayo, the farm was an opportunity for "ecological propaganda" and "total change."[26] His intentions were utopian in the sense proposed by Fisk's acknowledged intellectual mentor, Murray Bookchin (who is discussed briefly in Chapter 1).

The agenda of those activists with ties to the organic gardening movement, the third ecologist constituency, was less overtly political. But they, too, made a correlation between improving the soil and the quality of people's lives. The soft-spoken gardener Thomas Rosas understood that radical change would bring to Laredoans a proper diet, decent drinking water, and less dependency on chemicals.

These three varieties of ecologists—designers, labor organizers, and gardeners—understood their role in the community as conservators, "as," in Fisk's words, "a catalyst to get people to think about the boundaries of home." But rather than conceiving of social boundaries in a proprietary sense, Fisk in particular was most interested in identifying the social dimension of "natural forces." As the nominal leader of the ecologists, he planned first to put the buildings together entirely with scrap—materials recycled from the urban waste stream. The idea was to focus people's attention on "hidden resources"— not only industrial wastes such as fly-ash and recycled oil-pipe stems, but local (nontraditional) building materials such as caliche and buffle grass. Fisk's interests were more focused on how "product development effects broad scales of resource replacement" than on social spaces in any traditional, or architectural, sense of that term.[27] Simply put, his intention was to build a *social machine,* or to design a *social/biological process,* rather than making a static object, building, or space.

Fisk's interest in the design of biological processes had been developing for several years, but was given a jump-start by his discovery of a 1978 article by the chemist E. S. Lipinsky. Lipinsky and his colleagues at the Battelle Columbus Laboratories in Columbus, Ohio, had speculated upon the development of "energy farms" that might "integrate the production of fuels from biomass with the production of food and materials, to form adaptive systems."[28] At the time of Lipinsky's research in the late 1970's, the oil crisis had precipitated substantial federal funding in search of an alternative to fossil fuels. Lipinsky's insight, and his critique of conventional ethanol science, was that the conversion of biomass to fuel might be only one strategy in a more radical rethinking of agricultural production. Fisk eventually described such an adaptive mode of agricultural production as a "flexible farm." Although Fisk never seriously proposed that Blueprint Farm become the site of alternative fuel production, he was deeply influenced by Lipinsky's systems approach to agricultural production. The notion of "flexible farming" became the technological model for Fisk's intentions at Blueprint Farm.[29] Figure 4.1, a diagram that conceptualizes many of the systems ultimately proposed for the Laredo project, illustrates Fisk's initial proposal for "Future Farms of Texas."

The computer-generated drawing illustrated in Figure 4.1 was presented by CMPBS at its initial meeting with TDA to discuss the development of Blueprint Farm. It was Fisk's intention that the flexible farming system he had envisioned become an "interactive modeling procedure" that would be "responsive to decision making by others over time." The implications for democratic participation afforded by this proposal are significant indeed. Unfortunately, the computer-based model illustrated here was never developed by CMPBS. The immediate demands of producing the farm itself intervened. In lieu of the in-

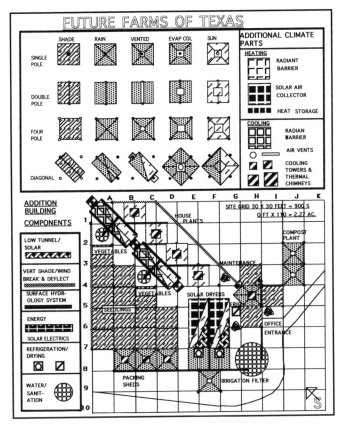

Figure 4.1. "Future Farms of Texas" by Pliny Fisk III, © 1988 CMPBS. Courtesy of the Center for Maximum Potential Building Systems. This illustration of flexible farming systems was Fisk's conceptual model for Blueprint Farm.

teractive computer model envisioned by Fisk, the physical model illustrated in Figure 5.1 was used in a limited "gaming" context with a few TDA officials to investigate alternative configurations of the technologies. In subsequent chapters, Chapter 5 in particular, we will see that the loss of this opportunity to democratize the conceptual planning of the farm had dire consequences.[30]

By his own admission, Pliny Fisk loves technology. His enthusiasm for the topic is never simply instrumental, but seems to arise from what the historian Arnold Pacey describes as the existential pleasures of interpreting the world through material engagement—the pleasure of making stuff that works. For Fisk, as for Pacey, applications of technology are a way of giving the world meaning.[31] Fisk argues that "technology exists inside other things, inside natural forces and people." In this sense, "technology is a translator," an interpreter of place, because it adapts culture to a specific ecological context, and, symmetrically, it adapts environment to cultural context. Fisk's goal for Blueprint Farm was to render the place "coexistent with natural forces." By "natural forces" Fisk intends the sun, the wind, and the rain, as well as the organic cycles of growth and decay, source and resource. He holds that "Natural forces transformed into information become a guide to the interpretation of artifacts." For Fisk, in language most unusual for a designer, "the issue here is to get humans in touch with nonspatial things, with resources." With similar unconventionality, he asserts that Blueprint Farm is not intended "to be friendly to people. It is not about accommodating humans."[32] Presumably he means that the interpretation of landscape is more important than either human comfort or material production.

Alvaro Lacayo reflected that the Meadows Foundation (the principal grantors of the project) was actively hoping to forge a union between the Israeli scientists and the Texas-based ecologists. In the forced marriage constructed by the Meadows Foundation, previously isolated networks were consciously woven together. Fisk is fond of noting that "we survive by networks." He means by this term, not only the political networks that "constantly realign information," but that "everything is connected; housing, farming, the river ... etc." But he also laments that "there is an expert mentality problem with networks." By "expert" he refers to the close guarding of proprietary or disciplinary interests. It seems that his objection is aimed at the distinction between alternative meanings of the term *network* itself. In the narrow political sense of the term, it describes interconnected human or institutional interests. In the broader sense that Fisk prefers, it describes relationships between humans and nonhumans that "cross the barriers between art and science."[33] It is in the former sense that Fisk's own authority most commonly came into question. Because he insisted upon integrating narrowly defined disciplinary networks in the making of Blueprint Farm, the experts of the fields in which he trespassed—principally those of his Israeli counterparts—constantly suspected his intentions. The result of Israeli suspicion was increasing marginalization of Fisk's position. In response, Fisk fashioned an identity for himself at the edge of institutional life where the potential for failure is worn as a badge of distinction. As Fisk was shoved further and further into the margin of the "exchange" with Israel, he began to see conspiracies everywhere. Not only were the "ultra-dogmas" of Israelis questioned, but vast corporate collusion was seen as the source of his compromised condition. The historian Reuven Brenner offers the possible hypothesis that the very real conflict between Fisk and the Israeli managers of Blueprint Farm may be understood as a case of rational "contract uncertainty."[34]

The ecologists intended a radical social change that would benefit both humans and nonhumans. The machines constructed by Fisk were an attempt to harness human practices to the cycles of nature. In the process, however, the ecologists (and Fisk in particular) proved insensitive to the interests of competing networks. By insisting upon the superiority of their own program, without *demonstrating* its relevance to potential detractors (the Israelis in particular), they ensured their own isolation. One cannot avoid the hypothesis that the subliminal interest of the ecologists was to fail, thereby demonstrating the uncompromised purity of their utopian program.[35] In this sense, the ecologists intended to leave a blueprint for others to follow, not to construct a sustained place. The interests of the ecologists were rigidly ideological.

Although the views of Laredo Junior College trustees and officials were diverse, there was an explicit agreement among them that the intent of Blueprint Farm was to "save the small farmer,"[36] "improve people's standard of living,"[37] and "make a profit."[38] There was also an implicit understanding among local experts that the Israelis would show them how to do these things—how to fix the way things were. Because the reports of Israeli capabilities were so impressive, and the conditions at Laredo so difficult, Dean Jacinto Juárez and others argued that Laredo was the perfect location for a demonstration of the new technol-

The Local Expert Network

ogy. If agricultural production could be increased tenfold in Laredo, as it had been in Israel, profitability could be returned to small farms. In 1996, well after the termination of the project, Dr. Jiménez lamented that "we thought the scientists would find a solution."[39]

The administration of Laredo Junior College understood the project in the terms of economic development. Finding a productive use for small family plots of land that had been subdivided time and again over the years could restore unproductive capital—meaning the land—to profitability. Dr. Héctor Jiménez conceptualized the farm as a "root" that would allow local families, who traveled seasonally as migrant farmworkers, to literally dig in. If the new technology could transform their diminished land assets to productivity, the impoverished workers could stay home to farm. For local experts references to the technologies employed at Blueprint Farm meant methods that would "maximize water resources." Because of the daily reminders of the ongoing drought, popular understanding of advanced agricultural technologies was linked to the Israeli trickle-irrigation system. For these locals the concept of drought control became synonymous with the concept of sustainability. Even after the termination of the project, none of the local experts involved in the production of the project could offer a definition of sustainability that was broader than the limited concept of resource conservation.

At the beginning of the project, local business and education leaders reasoned that if the land grant network was unwilling to change local conditions, a competing network might be willing to try. The logic of local experts, in attempting to enlist Hightower and the Jewish network in their cause, was that of hard technological determinism. The interest of local experts was in the importation of a technological fix.

The Land Grant Network　　From Jim Hightower's point of view, the land grant network consisted of six distinct institutions: Texas A&M (the state land grant university), the U.S. Agricultural Extension Service (AES), the corporate manufacturers of farm equipment and chemicals (such as John Deere and Monsanto), corporate food producers (such as Archer Daniels Midland), the U.S. Department of Agriculture, and his own agency, the Texas Department of Agriculture. To argue that Hightower's attack on the land grant network (the traditional ally of TDA) caused political turmoil in Texas would be an understatement. To argue that the land grant network was out to "get Hightower" would likewise be an understatement.

When Blueprint Farm was still in the early planning stages, Pat Roberts, a senior AES (Agricultural Extension Service) official, was assigned to monitor the project from his office at Texas A&M. Roberts was charged with "minimizing the damage" that Hightower might inflict upon the food industry. The perception inside AES was that the agency was in a defensive mode, trying to find out how minimal cooperation with the Hightower regime in Austin was possible in the face of "constant TDA antagonism." Some success, like the creation of the Pesticide Bureau to check Hightower's initiation of antichemical regulation, was marshaled in the legislature. However, such moments were, in Roberts's eyes, unfortunately few. "Getting Hightower" became a personal and agency obsession.

The Hightower network was perceived as symbolically dangerous to the land grant network because, in Pat Roberts's words, "Jim Hightower enhanced the myth that something is wrong with your food and water, and that he can fix it." To Roberts's AES colleagues, the technological arsenal with which Hightower claimed he would "fix" things included a bitter irony. AES engineers claimed that the so-called "Israeli" trickle-irrigation system was not, in fact, "invented" by Israelis. The claim of Israeli authorship was a slap in the face to agricultural engineers at Texas A&M because "Texas Aggies" had themselves developed the initial prototypes of that technology. To hear the constant suggestion that the Israelis were over here "helping us out" was more than AES officials could stomach.[40]

AES treated the Hightower regime in general, and Blueprint Farm in particular, as a wake-up call. When "normalcy" returned after the 1990 election, AES officials had already formulated a public response to the "myth," fabricated by leftists, of a "threat to the environment." This logic concluded that "the way to preserve the environment is to "increase productivity," thereby taking the pressure off the most valued resources. Pat Roberts speculated that, if the artificial political restraints imposed by ecologists were eliminated, Texas food producers could increase productivity 50 percent in five years. Roberts assumed that such impressive numbers—such "wealth potential"— would surely subvert any return to what he characterized as the grim prophecies fabricated by Hightower and the regime of sustainability.[41]

At the other end of the Texas A&M campus the philosopher Paul Thompson was trying to define the slippery term "sustainability." Although Thompson then held a joint faculty position in the Departments of Philosophy and Agriculture, he could only be considered a very marginal influence upon, or irritant to, the land grant network. His research does, however, document the emerging definitions of "sustainability" available to the land grant network and others.

In lieu of a single definition, Thompson's research reconstructed three competing dimensions of the term then in common use: the *economic,* the *ecological,* and the *social.*[42] Each of these dimensions attracts a shifting network of supporters and establishes ideological tensions among the three poles of the sustainability triangle. Simply stated, economic sustainability, or simple resource sufficiency, takes a supply-side approach to assuring an adequate supply of dwindling resources. "Yield-enhancing" strategies of increasing supply, through improved technologies or improved rates of production efficiency, are economic strategies that measure input-output ratios and the minimum natural resource capital to sustain human activity. Mainstream economists and industry generally support this concept. Economic sustainability was increasingly supported by AES in the 1990's.

Ecological sustainability, the second dimension, takes a demand-side view of the question. This position does not deny that the efficiency of production might be improved, or further optimized. However, ecological sustainability adopts a neo-Malthusian view based upon the concept of eco-scarcity—that ecosystems do determine the biological limits to supply. In other words, the control of human demand is likely to be more effective in extending resource availability than the maximization of ecosystem supply. Ecologists generally support this concept, but neoclassical economists and industry do not.

Social sustainability, or the question of social equity, the third dimension of the concept, considers the political question: what is *just* within the human community regarding the equalization of access to nonhuman resources? This position is well summarized by David Harvey, who argues that

> all ecological projects (and arguments) are simultaneously political-economic projects (and arguments) and [vice] versa. Ecological projects are never socially neutral any more than socio-political arguments are ecologically neutral.[43]

This dimension of sustainability is supported by the doctrines of social ecology, but not by those of deep ecology, neoclassical economics, or industry.

Simultaneously applying all three dimensions of sustainability to concrete situations is problematic, of course, because it implies a highly democratic discourse in which competing interests are rationally negotiated. It is highly unlikely, therefore, that meeting the demands of any one dimension—economic sustainability in particular—will meet the political and ecological demands of the concept as a whole. Writing in the era when the project was being developed, Thompson argued that the struggle to define the meaning of sustainability would be of profound political consequence. In 1998, however, Thompson expressed skepticism that the concept of sustainability would ever mobilize broad public support.[44]

Others, like the planner Scott Campbell and the English scholars Simon Guy and Graham Farmer, have constructed more complex definitions of sustainability. Campbell's triangular model of sustainability is both more political and more optimistic than Thompson's in that he relates the competing interests of economics, ecology, and equity in an explicitly discursive, or dialogic, structure that was previously only implicit in the concept.[45] Guy and Farmer, however, argue that the three competing principles identified by Thompson and Campbell—economy, ecology, and equity—are insufficient to account for the many ideological positions that have adopted the concept of sustainability as their own. In lieu of three competing principles, Guy and Farmer have identified six distinct "logics" that compete to associate the term "sustainability" with the limited agenda of a particular constituency.[46]

No matter how optimistic one is regarding the future viability of the concept, however, it is essential to recognize that, in the era that Blueprint Farm was emerging from the ground, the concept of sustainability was just emerging in public discourse. The more sophisticated definitions offered by Thompson, Campbell, Guy, and Farmer come well after the farm had already closed. The contested definition of the concept, then, was inscribed in the contested geographies formalized by the farm's DMZ.

So long as Hightower and the regime of sustainability remained in power (1980–1990), the intentions of the land grant network were threefold: first, to preserve pockets of bureaucratic authority within TDA itself; second, to unseat Hightower; and third, to employ the new language of sustainability for its own economic purposes. Pat Roberts and his colleagues recognized the power of this rhetoric to promote increased production in the name of resource sufficiency. By redescribing the concept of sustainability in their own interests, they could cause proposals intended to realize ecological and social sustainability to disappear. The interests of the land grant network were thus invested in strategies of defense and the control of language.

UNINHABITED

INTENTIONS

To conclude this chapter, I will argue that most of the producers of Blueprint Farm tragically misunderstood their constructed interests to be their intentions. Interpreted through the tradition of intentionality initiated by Heidegger, this misunderstanding places the producers of Blueprint Farm in a thoroughly modern situation. Figure 4.2 will help to summarize how the competing networks operating at Blueprint Farm sought satisfaction for their intentions in *actual objects* rather than *intentional objects*.

The significance of the directedness plotted in this figure is best understood in the relation of intentional objects to actual objects. These terms are, of course, Husserl's, and although they fail Heidegger's ontology, I still find them useful in this limited context. In each case plotted here, save one, networks tended to confuse their social and ecological intentions with the material objects themselves. Only the ecologists remained directed toward the development of social processes, or institutions, rather than reified objects. In the end, however, their intentions were frustrated, rather than satisfied, by social conditions and their own ideological interests. I intend the term "ideological" to mean only that the social aspirations of the ecologists did not extend to those in competing networks.

The confusion between interests and intentions, or between actual objects and intentional objects, that is reconstructed from the evidence gathered at Blueprint Farm is not an unusual condition in modern life. Rather, I'll hold that this confusion documents a generally American comportment toward technological objects as historically determinant. Technological determinists understand technology as having a historical trajectory that is independent of social processes. This view holds that the activities of society are determined by the objects that we have created. For example, it is now a popular belief that urban gang violence has a direct correspondence to violence witnessed on television. Merritt Roe Smith, an historian of technology, has argued that the American presumption of a privileged destiny is little more than a mass form of technological determinism.[47] We have imagined, until recently, that benevolent technologies will secure for us a unique standard of material well-being. Of course, contemporary Americans have now discovered the malevolent side of technology's unintended consequences. Domestic nuclear power, for example, has proven to be something quite different than we intended. But in both views of technology, the benevolent and the malevolent, or the technophilic and the technophobic, Americans have tended to understand that it is objects, rather than practices, which control history.

In contrast, the minority view of technological voluntarists holds that

THE NETWORK'S	COMPORTMENT	IS SATISFIED BY	INTENTIONAL OBJECTS,	NOT	ACTUAL OBJECTS.
The Hightower network	employed symbolic-instrumental research	to realize ⟶	social justice	not	grow big tomatoes.
The local expert network	sought a technological fix	to conceal ⟶	social processes (in black boxes)	not	to own a farm.
The Israeli network	employed commerce	to produce ⟶	material security	not	to become responsible for computers, pipes, valves, greenhouses, etc.
The land grant network	employed defensive and linguistic strategies	to maintain ⟶	economic hegemony	not	to produce architectural ruins.
The ecologist network	designed ideological blueprints	to satisfy ⟶	the operation of living machines that relate social institutions to natural processes	not	to be published in professional journals.

Figure 4.2. The directedness of networks toward objects.

technology has a historical trajectory that is entirely controlled by social processes. This view holds that the objects of our creation only *reflect* social conditions. For example, voluntarists might contend that parental guidance, with the help of the V-chip, will successfully socialize the television sets seen by millions of American adolescents and thus reduce urban crime.[48] Of course, the V-chip is another technological fix that supports the view of determinists. In this case, however, it provides at least the illusion of choice for voluntarists. My own view is that both of these positions, technological determinism and technological voluntarism, misunderstand the relationship between human intentions and the objects that we construct to satisfy them. I agree with the voluntarist notion that social processes can positively effect the economic, ecological, and political constitution of technological objects. However, I also agree with the determinist notion that once intentions are embodied in material form, they limit our choices to operate in a place. I am suggesting that the history of the spaces in which we find ourselves living creates opportunities for action and limits our choice as soon as we are there.

This argument leads to this chapter's final observation to be drawn from Blueprint Farm. Even when the farm was in full operation, the sense of place, or structure of feeling, that pervaded the farm was *temporariness*. For example, the worker housing that Fisk and his ecologist collaborators proposed was never built because the very idea of "worker housing" must have sounded like a communist conspiracy to their collaborators. As a result, no one, including the Israeli manager, ever *lived* on the farm. In this sense the farm was *uninhabited*. This notion of *habitation* reflects Martin Heidegger's understanding of intentionality. For him, a place is necessarily inhabited before it is literally constructed. Heidegger means by this, I presume, that one's intentions for Being-in-a-place precede planning, precede the construction of objects, and precede the physical occupation of transformed space. For Heidegger, what shows up as a place is shaped by our negotiated manner of inhabiting that place. My point here is that it is human practices, rather than reified objects, that satisfy intentions. The tragedy of Blueprint Farm is that its producers, with the exception of the ecologists, failed to recognize how technologies are literally inscribed in the economic, ecological, *and* political agreements that eventually show up as places.

It was the intention of the ecologists to build a *living machine*—a process that would relate human practices to the natural forces of the place.[49] The ecologists, however, failed to focus their coproducers' horizon of intentions into a unified campaign to live and farm differently. Their own ideological interests prevented them from focusing the intentions of others into a set of practices that was economically, ecologically, and socially coherent. Instead of a *living machine,* what did get built was a *black box*—an object intended to maximize output and minimize input, without concern for the social processes that lie inside.[50]

CHAPTER FIVE

TECHNOLOGICAL INTERVENTIONS

Most of the difficulties we have in understanding science and technology proceeds from our belief that space and time exist independently as an unshakable frame of reference *inside which* events and places would occur. This belief makes it impossible to understand how different spaces and different times may be produced *inside the networks* built to mobilize, cumulate and recombine the world.

Bruno Latour, *Science in Action,* p. 228

Understanding technology in the terms of intentionality is helpful, but in the end inadequate for my purpose. Distinguishing between the interests and intentions of the five networks at work on the farm tells us something about the competing assumptions that brought these technological networks into conflict, but it has not yet told us much about how power relations are altered by technological interventions. In this chapter I will, first, investigate the tradition of science and technology studies as a mode of interpreting the social construction of technological systems, and second, apply this literature to Blueprint Farm by interpreting one of the technologies employed by the ecologists. In all, fifteen nonconventional technologies were put in place on the farm. A detailed interpretation of all fifteen might be desirable to some readers, but would no doubt be tedious to most. So as not to burden readers with an overly detailed account, I have selected one technology for interpretation and have presented descriptive information about the remaining fourteen nontraditional technologies in the Appendix, "The Things Themselves."

Choosing to interpret the technologies of Blueprint Farm through the tradition of science and technology studies may seem a curious choice to the reader following Chapter 4, in which Martin Heidegger's ontology played such a large role. Heidegger is, after all, considered a principal critic of modern technology. However, I want to again distinguish between Heidegger's philosophy of relation that emerges from his early work, particularly *Being and Time,* and his later poetic work, particularly *The Question Concerning Technology.* Without launching into an exhaustive critique of Heidegger's philosophy of technology, I will simply state that I find it to be overly deterministic and therefore not helpful in our current situation. Heidegger seems to have concluded that modern technology has a trajectory in history that is independent of social action. Like Latour, I do not find this deterministic position to necessarily follow from the ontology of relations constructed in *Being and Time.* In fact, the reverse is the case. By his own account, Heidegger's ontology requires a practice-based interpretation of reality, yet his philosophy of technology was constructed, without any empirical data, from the confines of his study in the Black Forest. Paradoxically, Heidegger has not followed his own very sound advice. His philosophy of technology is based upon the distant observation of technological outcomes, not empirical study of technological practices. As a result, I have chosen to rely upon Heidegger's philosophy of relation and leave his philosophy of technology to the postmoderns.[1]

Fortunately, the lack of empirical evidence in Heidegger's late work is overcome by science and technology studies (STS) in general, and by Bruno Latour's *actor-network theory* in particular. The bulk of this chapter will rely upon STS authors to better understand how a technology is socially constructed to produce social space. Actor-network theory is, however, less concerned with how worlds are produced than in how they are *received* and *re-produced.* These concerns will be taken up in Chapters 6 and 7, respectively.

TRADITIONS IN

SCIENCE AND

TECHNOLOGY STUDIES Conventional architectural history commonly awards the authorship of large, complex projects involving hundreds of workers to a single architect. We understand, for example, the Salk Institute at La Jolla, California, to be a *work* of Louis Kahn, not the hundreds of other architects, engineers, contractors, and suppliers who contributed thought and labor to its construction. The weakness of understanding architecture as an individual work, or creation, is that the emphasis on authorship easily leads to a *great man* theory of history. Architectural invention and technological invention are popularly understood to be different things. In the popular imagination invention in architecture is understood to be a matter of rhetorical flourish conceived by a single designer. In this view the act of design is both asocial and immaterial. In contrast, invention in technology is popularly understood to be a matter of material process that requires the mobilization of complex human and nonhuman resources. In the view of science and technology studies, however, design is denied its privileged status as art making. All invention is understood to be both socially constructed *and* material. Without denying the very existence of rhetorical flourish in matters of architectural expression, I wish to focus this chapter on technological processes because that is how Blueprint Farm was perceived and conceived by its builders. Blueprint Farm was not conceived as a representation—except, perhaps, by Jim Hightower. Rather, it was conceived as an operational technology.

Social science has challenged the great man myth of history through the development of four traditions of inquiry. Each tradition demonstrates the social, rather than individual, construction of technology. These four traditions—*systems theory, social constructivist theory, critical theory,* and *actor-network theory*—will all have something to contribute to my interpretation of Blueprint Farm. To some degree I will graze pragmatically among them. In the end, however, it is the assumptions and methods of Latour's *actor-network theory* that will prove most helpful.

The first tradition in the field of science and technology studies is social constructivist theory. Wiebe E. Bijker and John Law, among others, have demonstrated the "contingent" nature of technological agreements.[2] They argue that in the contemporary economy, the contingent agreements between producers are designed in the flexible interests of capital. In this view, technoscience is understood simply as one "belief-system" among many.[3] Radical social constructivists refuse to grant science any epistemological privilege over other ways of understanding the world. In this argument, social constructivism is pure epistemological relativism. Within architectural discourse, Alberto Pérez-Gómez comes closest to the constructivist position by observing the "weak" and "mysterious" origins of technology.[4]

A relevant example of social constructivist analysis outside of architectural discourse is Ruth Cowan's case study of the refrigerator. In "How the Refrigerator Got Its Hum," Cowan traces the failed development of the gas-fueled absorption technology and the successful electrically fueled compression technology.[5] Most scientists recognize that, considered from a purely technical position, the gas technology has superior life-cycle characteristics. Not only do gas refrigerators consume less energy, but the simplicity of their construction virtually guarantees a prolonged life in return for the initial investment. However, Cowan's study makes clear that the electric technology won the market, not because it was a superior technology, nor even because of consumer preference, but because the companies that opted for the electrically operated technology were, first, better financed and, second, more committed to the electric utilities than to the consumer. Cowan's claim is that "the machine that was best from the point of view of the producer was not necessarily best from the point of view of the consumer."[6] Thus, power relations among competing interests are revealed to be the primary context for decision making. The lesson to be learned from this social constructivist analysis is that technology may be shaped more by the need to preserve social and political forms than by science or the problem defined by the producers themselves.

Second in the field of science and technology studies is the tradition of systems theory, pioneered by Thomas Hughes. Hughes holds that Thomas Edison, for example, was not the inventor of electricity so much as the actor who integrated economic and political systems with scientific techniques that were already available.[7] In this view Edison was the quintessential entrepreneur rather than the secretive alchemist. What Edison administered was not the *invention* of techniques, but rather the creation of social agreements that would institutionalize the production and reproduction of previously isolated technologies. Systems theory understands technological innovation, not in terms of objects, nor in terms of techniques, but in terms of the human agreements required to standardize production and thus assure reproduction. Hughes's emphasis upon systems, or networks of agreements, is methodologically related to actor-network theory, but epistemologically related to social constructivism.

The third tradition of science and technology studies is the critical theory of technology articulated by Andrew Feenberg. As a Marxist following in the tradition of the Frankfurt School, Feenberg is hardly surprising in rejecting what he refers to as the late Heidegger's "substantive theory of technology." It is also not surprising that he would simply ignore the social constructivist theory of technology because, in Feenberg's view, that literature simply ignores the unequal power relationships constructed by market-driven technological systems. Less predictable, however, is Feenberg's rejection of the determinism implicit in the traditional Marxist position toward technology and "civilizational change." In the wake of the Soviet failure, and the entirely instrumental use of technology by Soviet managers, Feenberg reverses the Marxist assumption that it is the state that can most effectively control the "ontological decision[s]" implicit in technological choice. In the place of direct state control of production from above, Feenberg argues for a "politics of technological transformation," or a "parliament of things," in which a radical participatory democracy in the design of technological systems is realized. Feenberg argues that a shift toward partici-patory democracy in the workplace would both reduce the "operational au-tonomy" of managers to dominate production in the name of efficiency and increase the ability of workers to realize their own creative potential. In this argument, Feenberg is most concerned with the selection of technologies as the selection of a way of life. He rejects, however, the "dilemma of development," in which the social concern for worker satisfaction is understood as a costly "trade-off" in the purely mathematical rationale of productive efficiency. Rather, Feenberg reasons that market-driven societies have simply suffered a conve-nient failure of imagination in not constructing alternative technological ra-tionales that are both efficient *and* life-enhancing.[8]

As a concrete example of how history might be otherwise, Feenberg relies upon what Terry Winograd and Fernando Flores describe as the "ontological designing" of computer systems.[9] In the design of computer software, these scientists understand their project to be, not the hierarchical control of data categories intended to command and replace workers, but providing tools in-tended to enhance the decision-making capabilities of workers operating in a collective setting. *Ontological designing* does not blindly accept the technologi-cal imperative imagined by capitalists hooked to the technological treadmill. Rather, it subverts instrumental logic on behalf of Being. It is no accident that Winograd and Flores's research is based upon the early Heidegger's ontology as articulated in *Being and Time*. It is likewise significant that Feenberg, as a Marxist, would recognize the salience of this proposal to his own democratic concerns. Feenberg's critical theory of technology thus builds a conceptual bridge to Latour's actor-network theory, in that they share ontological concerns derived from Heidegger's philosophy of relations.

Finally, the fourth tradition of science and technology studies is actor-net-work theory. Bruno Latour, the most well-known contributor to this position, has recognized that the great men of science, rather than being autonomous hero-inventors, are men propped up in the eyes of society by the material pro-cess of knowledge production.[10] For Latour, the facts claimed by "technoscience," far from being the discoveries of disinterested scientists, are demonstrated to be socially constructed by the material interests of the social networks that produce them. Once scientific facts become encased in the "black box" of peer review, the web of interested agreements that support the artifact cannot be effectively challenged from the outside. Scientific truth, for Latour, is neither the asocial objective Truth imagined by realists, nor the interchangeable social constructions imagined by relativists. Rather, Latour understands truth to be manifest in the relationships that emerge between humans and nonhumans—the relationships that *show up* between *quasi-objects*. This is not a compromise epistemology that reconciles positivism with relativism, but an ontological position that follows from Heidegger's philosophy of relations.[11]

To support his ontological interpretation of technoscience, Latour has conducted a series of empirical or practice-based studies using a method described by social scientists as "participant observation." In what is perhaps Latour's most famous use of this method, he spent two years working as a lowly lab technician at the Salk Institute in La Jolla, California. While he performed the mundane chores of a lab technician, Latour also observed the daily activities of scientists engaged in the production of "facts." Latour's interest in this investigation was not the architecture of Louis Kahn, nor even architecture as the social setting for scientific decision making, but rather the social production of facts themselves. Latour concluded that "scientific activity is not 'about nature,' it is a fierce fight to *construct* reality."[12] Reality construction, in the sense intended by Latour, suggests that the "facts" through which society interprets the world *appear* to be unconstructed by anyone. As a result, those constructors who remain veiled by the opaque networks of technoscience enjoy significant privilege in society—they are the unacknowledged bankers of truth.

Latour's project is to again make visible the politics embedded in science and the science embedded in politics. He argues, following Heidegger, that science is not "idea based," as traditional positivists have claimed; rather, it is "practice based."[13] The debate between these concepts that Latour wants to reconstruct is embodied in the question: how shall we "practice" technoscience? If technoscience is, at its heart, a political practice, how are we to interpret the politics of Blueprint Farm and the construction of sustainable technology as a socially subscribed concept? The method of conventional architectural criticism would be to interpret the artifact itself, as did the "enlightened observer" pictured in Chapter 1. Latour, however, requires an analysis of the social practices inscribed in the artifacts. If you want to understand how technologies are produced and reproduced in the world, reasons Latour, don't study what the practitioners of technoscience *say*, study what they *do!*

The sum of these four traditions of inquiry—systems theory, social constructivist theory, critical theory, and actor-network theory—is to erase any validity to the claim that autonomous individuals might be responsible for the "invention" of sustainable technologies. In this regard, science and technology studies will prove helpful in understanding the competitive and bureaucratic conditions encountered in the development of Blueprint Farm. We are left also, however, with the task of understanding the complex relationship between the construction of societies and technologies. If individuals do not invent technology, the reverse causality—that technologies invent society—is not necessarily true either. This returns the discussion to the question of technological determinism.[14]

Simply understood, technological determinism "makes it seem as though the end of the story was inevitable from the very beginning by projecting the abstract technical logic of the finished object back into the past as a cause of development."[15] Arguing from a social constructivist perspective, and against the doctrine of determinism, Donald MacKenzie and Judith Wajcman hold that "a new device merely opens a door; it does not compel one to enter."[16] Theirs is a voluntarist view of causality. They reason that individuals are compelled by the structure of human agreements, not the structure of objects. Heidegger's philosophy of technology, as I have previously argued, accepts the premise that modern technology has a trajectory that is independent of social action. It is determinism writ large. Actor-network theorists have, however, developed a third position that avoids both the determinism of Heidegger and the voluntarism of social constructivists like MacKenzie and Wajcman. They argue that human agreements and objects are engaged along a continuum of influence. Although the structure of human agreements is literally inscribed in the structure of artifacts, the structures of artifacts limit subsequent agreements. Latour, in particular, argues that human agreements do not exist independently of the practices physically embodied in artifacts. To distance ourselves from the material implications of human agreements, or to distance agreements from their material context, is only to reproduce the myth of autonomous science invented by traditional positivists. In Latour's view, objects *do* compel practices because practices are always already contained in artifacts. It is only the mask of a seemingly objective science that allows us to imagine that it is otherwise. The debate concerning causality cuts both ways.

My own position lies closest to Latour's—a position that recognizes the contingent power of artifacts, or what Merritt Roe Smith has described as "soft determinism."[17] On this account, technology shapes society, but is in turn shaped by society. But this proposition is always locally applied, meaning that the relative dominance of technology or society is situational. Another way to argue this point is to hold that *agreements may be inscribed in the structure of objects, but that all technological agreements also take place outside objects.*

To argue that objects compel practices is to argue that there are power relations embodied within artifacts. Although I will not suggest that Langdon Winner is an advocate of actor-network theory, his claim regarding the agency contained in artifacts supports that position:

The claim [is] that the machines, structure, and systems of modern material culture can be accurately judged not only for their contributions of efficiency and productivity, not merely for their positive and negative environmental side effects, but also for the ways in which they can embody specific forms of power and authority.[18]

Winner argues that there are two ways in which artifacts can contain political properties: first are designs that become ways of settling issues within a community. An example of this first category of political objects would be the bridges designed by Robert Moses, the dictatorial planner of New York City in the 1950's. These bridges were designed with low overhead passage dimensions so that they would exclude buses—and thus the poor who rode them—from wealthy suburban enclaves. In this case the issue of racial segregation was settled by design.

The second type of artifact that contains political properties is artifacts compatible with only certain kinds of social relationships. Atomic power is such an example. The potential danger imposed upon society by atomic reactors permits absolutely no flexibility in the authoritarian human relationships that are required to operate them safely. In other words, an absolute social hierarchy is compelled by the operation of the artifact. This argument is less that of a technological determinist than it is of someone, like David Nye, who recognizes that each regime of energy production—whether muscle power, steam power, electric power, or atomic power—reflects social choices about human power relationships.[19] Extending similar logic, Winner suggests that the development of energy derived from renewable resources may produce "intrinsically democratic, egalitarian, communitarian technologies."[20] Such a proposal is related to, but in disagreement with, Andrew Feenberg's notion of "subversive rationality." Feenberg argues that "authoritarian social hierarchy is truly a contingent dimension of technical progress."[21] In Winner's view, hierarchical relations are subverted by selecting other technologies. In Feenberg's view, hierarchical relations are subverted by using the same technologies differently. My own view is again centrist in that I find Winner's abstract logic to be overly determinist and Feenberg's abstract logic to be overly voluntarist. This speculation concerning subversive technology selection has obvious implications for an interpretation of Blueprint Farm, a speculation to which we will return. We shall see, however, that in the concrete conditions of the farm, Winner's logic proves more helpful.

If technology is neither invented by great men, nor entirely determinant of social trajectory, we can at least conclude that technologies, no matter what their social context, "might have been otherwise."[22] Most science and technology scholars are in agreement that there is no inherent structure that leads technologies to institutional closure in one direction or the other. The only things—other than the limitations of physics—that cause technologies to take the form that they do are the interests of those who construct them and the conditions of nature upon which they work.[23]

A final reference to the social constructivist model for the development of technology will be helpful before applying these constructs to the specific conditions of Blueprint Farm. Positivist science assumes a linear model of research and development of technological artifacts. In contrast, constructivists favor a "multidirectional model."[24] This model holds that certain directions in technological development die off and others are economically reinforced as members of the society come to share a set of meanings, or benefits, attached to the artifact in question. The point to be emphasized here is that the moment at which the artifact becomes socially "stabilized" is commonly confused with the moment of "invention." "Different interpretations of nature are available to scientists and hence ... nature alone does not provide a determinant outcome to scientific debate."[25] The same logic can be applied to technology and to sustainable architecture. In other words, there is "interpretive flexibility" attached to any artifact—it might be designed in another way.

For the purposes of this study, we might repackage the concept of "invention" as *the resolution of cultural conflict.* T. J. Pinch and W. E. Bijker have suggested that

Closure in technology involves the stabilization of an artifact and the "disappearance" of problems. To close a technological "controversy," one need not *solve* the problems in the common sense of that word. The key point is whether the relevant social groups *see* the problem as being solved.[26]

Of course, an emergent technology, be it conventional or sustainable, can reach "closure" through an agreement to ignore, or suppress, as easily as through an agreement to incorporate that technology into the fabric of daily life. I argue that Blueprint Farm reached closure through an agreement to suppress.

The five technological networks discussed in Chapter 4— the local expert network, the Israeli network, the Hightower network, the land grant network, and the ecologist network—each pursued separate interests and intentions on

MAKING PROBLEMS
GO AWAY

the farm. These were embodied, to one extent or another, in the fifteen "sustainable" technologies developed there. The Jewish network promoted four and the ecologist network promoted eleven of these fifteen competing technologies. As the project took shape, however, there was no common vision of sustainable architecture, agriculture, or technology that bound these five competing networks together. None of the individual technological networks was able to mobilize the human and nonhuman resources required to sustain development. In the battle for the imaginative supremacy to define reality, and the politically useful concept of sustainability, there were no victors.

The "problems" inscribed in Blueprint Farm were made to finally "disappear" by the executive order of Rick Perry, the Republican commissioner of agriculture who defeated Jim Hightower in 1990. But the Perry administration, and its supporters in the land grant network, were not the sole reasons for the "closure" of the Laredo experiment in sustainable technology. Internal conflicts, as much as external reprisals, were the reason that some participants saw their problem as being solved by closure of the project. To argue this point more clearly, it will be helpful to return to the five networks invested in the production of Blueprint Farm.

I have argued before that the local expert network—Laredo Junior College administrators and local business leaders—can be understood as hard technological determinists. As the passive receivers of autonomous artifacts, they did not expect to have to contribute to knowledge production, induce social change, or subsidize the economic operation of the imported objects.[27] When push came to shove, and the Texas Department of Agriculture began making economic demands upon Laredo Junior College, the problem definition shifted. Rather than a focus on the impoverished lives of migrant workers—the population that Hightower and Jacinto Juárez identified as the principal beneficiaries of the project—the problem that came to compel the college trustees was the economic albatross around their necks. After Hightower's election defeat, the president of Laredo Junior College was only too happy to announce that the project had been "a success" and would now be closed. When the problem definition changed, so did the solution.

The Israelis, with support from the local Jewish community, were trying to achieve what Thomas Edison had achieved more than a century earlier. Their project was to integrate political and economic systems with an existing, but underutilized, technique. That existing technique was the trickle-irrigation technology that the Israelis had appropriated from the Agricultural Extension Service, based at Texas A&M, some years earlier. Although the technique itself was, ironically, at home in Texas, the political and economic systems that the Israelis imported along with the pipes, computers, and greenhouses were not at all at home. The integrated system that the Israelis had developed in the highly managed and state-supported economy of the kibbutz proved unrelated to local political and economic realities. The kibbutz as a socioeconomic model is, of course, a socialist's dream. But the kibbutz economy thrives in Israel because of cooperative management conditions, a willingness to reinvest a large percentage of capital in productive assets, and a high degree of social cohesiveness. None of these conditions was available in Laredo. That the kibbutzniks were in America to transform their socially derived technological system into an export product (in a highly competitive capitalist market) is yet another irony of this case. As the Israelis began to surmise that technological transliteration was going to be difficult, they lowered their expectations. When the opportunity arose to transplant their resources into the for-profit sector, further down the Rio Grande Valley, their "problem" was solved and they departed.

The Texas Department of Agriculture, in spite of the radical agenda forced upon it by the Hightower network, managed to operate like an entrenched bureaucracy. Bureaucrats held over from the pre-Hightower, presustainability regime operated to enforce traditional categories of production, rather than to push the concerns of sustainable science or the political problem defined by their leader. The need to preserve conventional social and political form at the bottom of the hierarchy subverted the most progressive ecological rhetoric issued at the top of the hierarchy. The low-level Texas-Israel Exchange administrator who became the nemesis of designer Pliny Fisk III was far more concerned with maintaining the social form of established accounting procedures than in forging a new relationship among politics, economy, and technology.[28] Her problem "disappeared" only when Fisk and the advocates of sustainability disappeared.

The land grant network, from behind the scenes, operated in this case as did the makers of the electric refrigerator. Like General Electric, Kelvinator, and Westinghouse in Ruth Cowan's study, the land grant network was vastly better financed than its ecologist competitors, and like these giant corporations, the land grant network was more committed to corporate interests than to those of the consumer. Unlike the case in which the gas-absorption refrigerator was suppressed, however, legions of scientists are not prepared to assert the superiority of solar refrigeration, wind-towers, or vegetative water filters over the conventional industrial technologies supported by the land grant network. Because these alternative technologies were never fully developed, the "problems" of the land grant network never fully appeared. The land grant network was successful in stabilizing these technologies in the minds of Texas farmers as marginal, and unsuccessful, experiments. The closure of Blueprint Farm was clearly in the interests of the land grant network.

By this count, four of the five networks invested in the production of Blueprint Farm found more problems to disappear by the closure of the farm than by its continued development. In this sense, the closure of the technological controversy inscribed in Blueprint Farm meant the reinvention of conventional technoscience. The ecologists, of course, were the odd network out. It was, as Langdon Winner has proposed, their goal to "develop intrinsically democratic, egalitarian, communitarian technologies."[29] There was, however, nothing about the closure of the farm that could be mistaken for the production of these conditions.

In the reconstruction of this case, it has been tempting (for some) to present the designer Pliny Fisk III as a repressed "great man," or alternately, as a local prophet ranting in the desert. But in the discussion above, I have argued that genius-inventors are, at best, social projections. In the systems theory tradition of science and technology studies, "inventors" are those holistic entrepreneurs, like Edison, who act to integrate economic and political concepts with emergent techniques of production. Edison's invention was not electricity, but a system of management and finance. Some might suggest that the critical difference between Edison and Fisk is the latter's lack of management and financial skills. Such a claim, reasonable as it may be in the face of TDA's accusation of financial mismanagement by Fisk, would, however, miss the difference between the contexts into which these men cast their "inventions." Where Edison invented into a fluid world stimulated by the economic potential of an emerging idea, Fisk invented into a world dominated by the hegemonic economic interests of the land grant network and entrenched power producers. Sustainable technologies, if they worked, would pose a significant challenge to the economic interests of Edison's heirs. I am not constructing a vague conspiracy theory that Texas oil producers, say, conspired to stop the demonstration of sustainable technologies emerging at Laredo. I am, however, suggesting that in the current global economy, it is not likely that Fisk will find the kind of management and financial support for his vision that Edison found readily available for his. It is more likely that Fisk's inventions—like his solar refrigerator—will follow the path of the gas refrigerator documented by Cowan.

The substantive question that remains is: were the suppressed technologies proposed by the ecologists somehow more democratic, egalitarian, or communitarian than the large industrial systems defended by the land grant network? One tragedy of the project is that we cannot know the answer to this question without evidence. And, as Latour has helped us to understand, the evidence that is needed is not the objects themselves so much as the human practices inscribed in them. Without an operational farm—a place where people work every day—it is simply not possible to *know* in the terms required by this inquiry.

It is, of course, possible to speculate, or hypothesize, how these technologies *might* operate on the basis of their limited development and use. It will be helpful, then, to briefly examine one of the technologies—say, wind-towers—as a concrete example of the ecologists' intentions. An examination of straw-bale walls, constructed wetlands, wind power, solar water heating, solar refrigeration, or solid-waste composting might be equally valuable. I have selected wind-towers, however, for at least three reasons: first, they are generally less familiar to readers; second, they are such a prominent part of the project's architecture; and third, like the project itself, this system never really worked. To understand the political implications of wind-towers as a cooling technology, it will be necessary to compare them to the conventional technology proposed by other architects in a competing design proposal for the farm.[30] That conventional cooling technology was, of course, compression air-conditioning.

Thomas Hughes has argued that as technological systems grow larger and more complex they tend to shape society more than they are shaped by society.[31] In this case, Hughes's insight is of primary importance in understanding the distinction between mature industrial and emergent sustainable technologies. The conventional industrial technology, compression air-conditioning, is ubiquitous in Texas. It is a huge industry dominated by three or four highly competitive industrial manufacturers. As the industry has matured, the variations in technology design have diminished dramatically so that product differentiation is largely a matter of service, packaging, and distribution. To all but the mechanical engineer, air-conditioning has become a

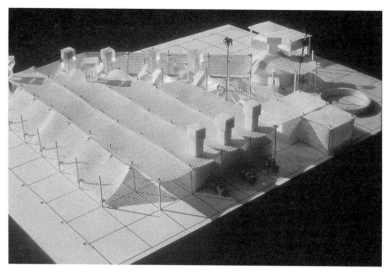

Figure 5.1. Model study of Blueprint Farm, by Pliny Fisk III, © 1988 CMPBS. Courtesy of the Center for Maximum Potential Building Systems.

classic "black box" that is rated in tons of output capacity.[32] One need not know how the contents of the black box transform conditions so long as the box continues to operate. These black boxes sit on rooftops, in closets, and behind shrubbery. I will argue that it is appropriate to understand conventional air-conditioning as an *appliance*—a device constructed in an assembly line at a distant location, imported to a local place, and plugged into the universal power grid.

My point here is that the importation of technologically opaque appliances contributes first to the *fetishization of commodities* and, second, to the *universalization of local space*. Both characteristics lead to social conditions that I must characterize as inherently undemocratic. By commodity fetishism, I refer to the abstraction of the appliances from the lives of those who produced them in the first place. In classical Marxist terms, it is money that acts as the neutral barrier between producers and consumers. Without money, we could not be so distant from, yet dependent upon, those producers whose aspirations remain so totally opaque to us. For example, in contemporary society, it seems almost absurd for me to think about the person who sewed the shirt that is now on my back. Should I consider the conditions under which she worked to provide for my comfort? Should I ask what she needs to satisfy her aspirations in life? As with the production and purchase of my shirt, the abstraction of the black box of air-conditioning both rewards us and protects us from knowing or caring about the conditions of its makers. But the key to human abstraction is not simply money, it is space. Or, put another way, money is an essentially spatial concept—it constructs social distance. If such black boxes were produced locally, in Laredo, the degree of alienation between producers and consumers would be somewhat mediated. The significant point here is that local workers visibly produced the wind-towers of Blueprint Farm. This technology was physically constructed in situ by members of a geographic community. It is the *visibility* of the conditions under which neighbors work that demands democratic and egalitarian practices. It is difficult, although not impossible, to abstract the lives with which we share space on a daily basis. Henri Lefebvre supports this point by arguing that "Productive operations tend in the main to cover their tracks." We polish, stain, and remove the scaffolding after construction is completed. In this way, products "detach themselves from productive labour." Modern construction, then, is "forgetful," or it is a "mystification" of productive labor "that makes possible the fetishism of commodities: the fact that commodities imply certain social relationships whose misapprehension they also ensure."[33]

My proposal for *visible* technology has been amplified by the landscape architect Rob Thayer.[34] He has argued in favor of *transparent,* as opposed to *opaque,* technology. In this distinction, Thayer has recognized the "cognitive dissonance" between how landscapes *appear* and what we *know* about them. Thayer argues that the increasing opacity of postmodern technology obscures the core operation of the artifact in favor of a constructed, and frequently unrelated, surface appearance. In the case of the wind-towers constructed at Blueprint Farm, however, not only can observers decode their operation visually, but the visual information is subsequently supported by the haptic experience of coolness. Transparent technologies thus make abstract knowledge concrete and particular, rather than secreting it away inside a black box.

My second observation is that air-conditioning acts to universalize local space. It does so by reducing the quality of experienced coolness to a universalized equivalency measured in tons.[35] Rather than utilize local sources of coolness—the breezes, the river, locally available minerals, or the ground itself—mechanical air-conditioning acts to convert all spaces to a mathematical, and universal, measurement of volume. While some might argue that such homogenization of space is radically democratic, in that all space is equalized, Hughes for one would argue that such technological shaping of social space is, in the final analysis, totalitarian. The wind-towers proposed by Fisk are, of course, the antithesis of mathematized space. They depend upon the understanding of very particular conditions. This technology is democratic in principle because it requires local consideration and distribution of available natural forces. The operation of wind-towers, unlike the operation of compression cooling, does not demand that local space be linked to, and made dependent upon, the centralized energy network tied to a distant "center of calculation."[36] Rather, local producers will have to participate in a public discourse in order to determine who has access to the wind and the sun. Public discourse is required because the flow of these natural forces is not (yet) subject to conventional concepts of property rights. The distributive justice of natural forces is necessarily directly democratic because the conditions of their production are local and particular. This is a political condition that Feenberg would, no doubt, find subversive and hopeful.

With these observations in mind, I will refer to the nonconventional technologies developed at Blueprint Farm as *regenerative* technologies rather than as *sustainable* technologies. The further definition of this term (which is, as I noted in Chapter 1, borrowed from the landscape architect John Lyle) will be undertaken in Chapter 8. I want to stress here, however, that *regenerative* technology should not be understood as synonymous with *passive* technology or as some veiled Luddite strategy. In the current discussion, regenerative technology adds to the economic, ecological, and social dimensions of sustainable technology discussed in Chapter 4 in two ways: First, regenerative technologies are socially visible and politically transparent in the manner discussed

Figure 5.2. Wind-tower about to be lifted into place, photograph by Pliny Fisk III, © 1989 CMPBS. Courtesy of the Center for Maximum Potential Building Systems.

above. Second, where sustainable technologies require only that the status quo of production/consumption be attained, regenerative technologies require a net increase in life-enhancing conditions. These conditions do not, as Feenberg insists, prohibit highly sophisticated material processes.

Unfortunately, this particular example of regenerative technology—wind-towers—did not operate well enough at Blueprint Farm to win supporters and stabilize a system that might be reproduced elsewhere. The reasons behind this failure are technically interesting, but not particularly salient to the current discussion.[37] It will suffice to say that others have had significant recent success with similar regenerative technologies in hot-arid climates. In spite of the inconclusive nature of the evidence, however, the political claims of the ecologists have merit. Without requiring the reader to endure a step-by-step analysis of all eleven localized technologies proposed by the ecolo-

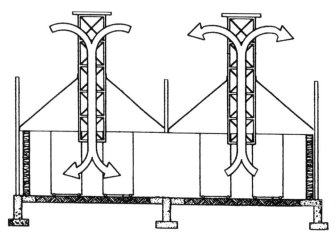

Figure 5.3. Diagram of updraft and downdraft wind-towers. Redrawn from original by CMPBS.

gists, I will argue that such localized systems of production point the way toward participatory democracy and egalitarian places. My point is that the allocation of natural forces, which are experienced equally by local decision makers, demands the kind of messy, public discourse that is fundamentally democratic. Such dialogic relationships are in distinct contrast to the absolute hierarchical relationships demanded by atomic power, or the corporate hierarchies seemingly demanded by the production and distribution of fossil fuels. Latour might argue that the dialogic relationships required by the production of regenerative technologies are literally inscribed in the structure of locally produced artifacts. If such a claim could hold up to rigorous scrutiny, the regenerative technologies proposed by the ecologists at Blueprint Farm could be said to be, as Winner would have it, both "inventions which become ways of settling issues within a community" and "inherently political." Technology design in this sense would be understood, as Sal Restivo would have it, as "a program for freedom and liberty in everyday life."[38] On the basis of the evidence provided by this case study, however, no such claim can be made with any certainty. The tragic and premature closure of Blueprint Farm prohibits an analysis of the practices that may or may not be inscribed in the artifacts themselves.

What can be said with some certainty about wind-towers is that they briefly opened new conceptual and literal space for emergent social practices. Even if for a short time, a few people imagined, lived, and farmed differently. By extending this logic one might argue the same for straw-bale walls and for the other nine regenerative technologies deployed on the farm by the ecologists. New networks of life-enhancing practices were cast. The degree to which these nets captured new spaces is, of course, the degree to which they can claim to be regenerative. Chapter 6 will consider how the artifacts of Blueprint Farm were received and Chapter 7 will consider how, if at all, they regenerated themselves in other spaces.

The focus of this chapter has been on the production of technologies and the modification of power relations implicit in technological interventions. To end with an inconclusive comment about the social power of artifacts would, however, conceal a principal irony of the case. According to Jim Hightower and Jacinto Juárez the primary beneficiaries of the technologies put in place at Blueprint Farm were to have been the agricultural workers of la Frontera Chica who have been displaced by industrialized farming. *The irony of the efforts to democratize agricultural production in the Rio Grande Valley is that no one bothered to ask the displaced workers what they wanted.*

DEMOCRACY AND PARTICIPATION

I have already argued that the local expert network—Laredo Junior College administrators and local business leaders—can be characterized as technological determinists because they expected autonomous technology to solve the problem of displaced labor. With a bit more hindsight I will finally argue that the Hightower network, the Jewish network, and even the ecologist network were affected by the same determinist expectation. Although the ecologists tried again and again to implement outreach programs within the local community, I hold that these were efforts to reproduce their own ideological interests, not efforts to understand what tools the displaced workers wanted to modify their own situation. Even before CMPBS had begun the design of Blueprint Farm, Pliny Fisk III had envisioned an "interactive modeling procedure" that is illustrated in Figure 4.1, Future Farms of Texas. Tragically, this computer-based game of design was never developed by CMPBS. There is no evidence to suggest, however, that Fisk ever envisioned that the farmers of la Frontera Chica would themselves participate in the farm's design.[39] Among thousands of pages of documentation and twelve in-depth interviews, there was not a single reference that documented any of the producers of Blueprint Farm asking local farmworkers what they thought should be done.

If, as Thomas Hughes and Bruno Latour have argued, technology is largely a matter of human agreements, then we can understand the closure of Blueprint Farm as an ironic and unintended agreement among a disparate group of technological determinists to suppress the self-determination of farmworkers. Although the regenerative technologies produced at Blueprint Farm have great potential to democratize local space, it is highly deterministic and authoritarian to imagine that political conditions might be positively changed without the active participation of those whose lives are most directly affected. As Feenberg has summarized the modern situation:

The reconciliation of legitimacy and efficiency in the democratic state is the modern utopia par excellence, nowhere so far fully realized. The reason for the difficulty lies in the *contradiction of participation and expertise,* the two foundations of the system.[40]

Considered through Feenberg's lens, Blueprint Farm is a classically modern confrontation between the simultaneous demands for *participation* and *expertise* in the development of life-enhancing technologies. I'll conclude this chapter by arguing that objects *do* compel practices. However, locals must be visibly engaged in the construction of objects and agreements if practices are to be reproduced. Without unity of conception and execution, no technology will be socially sustained.

CHAPTER SIX

RECEPTION

...the historical essence of an artwork cannot be elucidated by examining its production or by simply describing it. Rather, literature should be treated as a dialectic
process of production and *reception.*

Robert Holub, *Reception Theory: A Critical Introduction,* p. 2

To understand how the technologies constructed at Blueprint Farm are related to the social construction of place, it is not sufficient to consider, as we
did in Chapter 4, only the intentions of those who were invested in the farm's
production. Nor will sufficient understanding be gained by observing the artifacts themselves as we did in Chapter 5. If we are to understand the social
production of Blueprint Farm, and its constituent technologies, we must also
understand how the project was *received.* That is the concern of this chapter.
Each view of the place—each reception—provides a different kind of understanding. The relationship between *intention* and *reception* is, however, not a
neat dialectic pair. It gets messy because those who had intentions for the
place, its producers, have also become its audience, its receivers. This shift in
the status of particular people, or entire networks, is not simply a before-and-
after-the-fact categorization. It is more an observation about power. Those

who perceive themselves as *receivers* impart power to others. In this case there was nobody who did not occasionally envision herself as being on the receiving end—as being an *object* of production rather than a *subject.*[1]

This chapter will first draw upon the literature of *reception theory* as a mode of interpreting works. That literature will provide a ground from which to, second, examine the competing receptions given to the farm. As we shall see, the categories of reception are related to, but not quite the same as, the categories of production. I'll then conclude this chapter by categorizing the differing receptions to the farm as three interpretive paradigms. The dynamics among these categories of interpretation will prove helpful in considering the problem of *reproduction,* discussed in Chapter 7.

RECEPTION THEORY   In the discourse on modern architecture, human *works* are typically interpreted through a process of discovering and demonstrating their historic contexts and origins. Those critics who operate under modern assumptions argue that historical interpretation is possible, not only through understanding the conscious intentions of the maker, but also through understanding the "unconscious or structural features that escape the makers' attention or understanding."[2] Many have referred to this as the "depth model" of meaning. The emphasis of modernist interpretation has thus been upon the historic relationship between the author and the work. In the contemporary discourse on literary theory, however, the German critic Hans Robert Jauss has argued that works cannot be interpreted by merely elucidating the intentions of makers, examining the structural features of the work's production, or simply describing the work in question. Rather, Jauss holds that the interpretation of works is a dialectic process of understanding both production *and* reception. In this sense, *Rezeptionstheorie,* as it is called in German, can be associated with the general postmodern emphasis, not upon the author and the work, but upon the text and the reader.[3]

Jauss explicitly refers to Thomas Kuhn's notion of a "paradigm shift" as the foundation for his postmodern proposal for a radically democratic mode of interpretation. While there is well-documented danger in the application of literary theory to the interpretation of architecture—which is material rather than textual—Jauss's insights are helpful in expanding the critique of modern technology provided by the disciplines of philosophy and sociology that are cited above. In the United States, reader-response criticism is a related position, but distinctly different from the Marxist assumptions of those who developed reception theory at the University of Constance in the late 1960's. Other critical theorists have more recently extended Jauss's logic beyond literature to art and architecture.[4] For all of these reception theorists, the writer of the text, or the designer of the place, is less in control of its interpretation than we typically assume. Reception theorists understand interpretation to be an active, rather than passive, practice—one in which the interests of the receiver are exercised.

In his study of reception theory, the American literary critic Robert Holub insists that the constant reinterpretation required by the social continuum of reception permits the construction of historicity that relates past meaning to present conditions. It is, of course, the insistence upon the historical interpretation of texts and objects that identifies reception theory as a Marxist, or critical theory, position. What distinguishes Jauss from previous critical theorists is his interest in the reconciliation of the historical interpretation of objects with purely formal concerns. He accomplishes this task by adopting certain aspects of hermeneutic interpretation that relate the object of interpretation to the horizon of expectation experienced by the viewer. In this sense, Jauss's mode of interpretation is more dialogic than dialectic. In other words, his interest is in the resolution of competing interpretations rather than in unmasking the oppositions that distinguish competing interpretations of reality. This positive attitude toward the interpretation of works associates Jauss and Holub with the brand of critical theory inhabited by Kenneth Frampton that was introduced in Chapter 1.

Consistent with such a hermeneutic approach, Jauss, like Frampton, takes particular exception to the "aesthetics of negativity" promoted by Theodor Adorno, a principal founder of critical aesthetic theory. Jauss argues that any recipe for an aesthetic avant-garde that is purposefully exclusive—in that it denies identification with the local social condition—cannot hope to develop a new scheme of social praxis.[5] Like that of the Marxist geographer Henri Lefebvre, Jauss's project is one of "dis-alienation." In Lefebvre's view, the Marxist theory of alienation is relevant not only to the question of labor, but to ontological problems as well. He argues that the alienation typical of modern social practices tragically extends to our relationship to the natural cycles of places, "it encompasses life in its entirety."[6] Rather than exploit modern alienation for its ability to unmask hidden interests, the revised Marxist position promoted by Lefebvre and Jauss proposes a positive schema that might contribute to the development of common life-enhancing praxis. This critique of critical theory from the inside will, in both my summary of this chapter and in the conclusion of this study, play an important role in my own attempt to reconcile Frampton's proposal for critical regionalism and the postmodern proposal for a reconciliation with nature.

In Robert Holub's articulation of reception theory, he whittles Jauss's text down to a few essential propositions. Principal among these is that "a given paradigm creates both the technique for interpretation and the object to be interpreted."[7] This may be only another way of insisting, as Heidegger has, that an object (or place) is inhabited before it is built. In other words, our manner of understanding, or of interpreting, reality precedes the realities that we literally construct—places spring up in response to our manner of inhabiting space. That the producers of Blueprint Farm intended a new paradigm of inhabitation is perhaps too obvious to state. The ecologists never tired of demanding that people live differently. In Henri Lefebvre's terms the farm was an attempt, at least for the ecologists, to write a new chapter in the history of space. It is in this sense that the Kuhnian concept of the "paradigm shift" will be helpful in understanding the reception, rejection, and reproduction of the project.

In a separate study, which examines the reception of reception theory itself, Holub applies Jauss's insights to the importation of ideas from one culture to another. His conclusion is that "what matters most in the appropriation of a theory from another country is how it fits with an already established constellation in the importing country."[8] Holub's conclusion easily lends itself to understanding the importation and reception of the concept of sustainability in la Frontera Chica. Although this region has not been a separate country since its one year of independence in 1840, the history of local space documented in Chapter 3 clearly describes it as a culture unlikely to receive the organic dreams of the ecologists with much enthusiasm.

MIXED RECEPTIONS

This section investigates the reception provided the farm by seven distinct groups—Israelis, local Jews, local experts, local small farmers, the land grant network, Texas architects, and the ecologists. These interpretive groups are related to the *technological networks* discussed in Chapters 4 and 5, but I have sliced the data differently. In this chapter I have distinguished local Jews from Israelis, local farmers from local experts; I have added Texas architects as an interpretive group; and I have deleted the Hightower network. The distinction between local small farmers and local experts is significant because it recognizes that, although small farmers were expected to be the principal receivers of the project, their visibility in the dialogue was blocked by those local experts who presumed to speak on their behalf. I have included Texas architects in the roster of receivers because they provide an interpretation that is helpful in placing the ecologists in a professional context. With similar pragmatism, I have deleted any reference to the reception of the project by the Hightower network only because they were rarely on the scene and there is little data that is of help. Although I would like to make the case that *interpretive networks* are the same as *technological networks,* such a template for analysis doesn't seem to help my own project of interpretation. So, pragmatically,

I will leave the reconciliation of technological and interpretive networks for another project.

Like the place itself, the Israelis drew a mixed reception. The very real conflicts that erupted on the site were understood by some to be cultural—a lethal mix of *mañana* and the *sabras,* the "do it tomorrow" attitude of Hispanic locals and the compulsive "do it now" attitude of the Israelis. But, in general, the Israelis were well received as a hardworking, yet skeptical, lot. Most locals granted them a cautious respect. Some, however, especially the ecologists who were affected negatively by the Israeli presence, found them to be "arrogant" and distant. As receivers of the farm, the Israelis were unhappy with the lack of response by local farmers and totally disdainful of what appeared to them as the "poppycock" schemes of the ecologists. If they viewed themselves as scientists, they viewed the ecologists as alchemists—craft workers deluded by a mystical connection to the land.[9] They received the plans of the alchemists with competitive contempt. The low-tech processes of the green Texans only baffled the Israeli "scientists" and threatened their marketing investment. They received their own efforts at international trade with disappointment.

Local Jews interpreted conditions differently than their Israeli cousins. What mattered to them was the opportunity to mediate the cultural stereotypes that had so affected their own lives in the border region. They received the farm as an opportunity for cross-cultural conflict resolution, not a marketing opportunity, or as an experiment in alternative technology transfer. In contrast to the profit-driven reception of the Israelis, local Jews received the artifacts of Blueprint Farm as icons of goodwill. So long as their own status in the local community benefited from the destruction of old stereotypes, they were happy. The project was evidence that Jews, too, dig in the dirt and mark the ground. It was a way to prove that, as Vera Sassoon put it, "we are more than moneylenders."[10] That Israeli science beat Texans at their own game was a source of contained pride.

Local small farmers also received the project with a variety of interpretations. Those not invested in the production of the farm accepted it, not with experimental anticipation, but with a serious case of agri-phobia—a fear of agriculture as a way of life. The negative social history of farming in Laredo discussed in Chapter 3, combined with the seemingly positive trajectory of international trade, has made farming an almost invisible, or suppressed, social practice—especially among the young. Among older Laredoans, those who retain a memory of its agricultural past, the opposite was true. Older locals suffered from technophobia—an inability to identify with the alien, unstable, or "high-tech" gadgets being developed there.[11] It is tempting to speculate upon how these older small farmers, who relate positively to their agrarian traditions, would have received a recuperation of those communal landholding practices briefly discussed in Chapter 3. Unfortunately, the question cannot even be investigated because those who controlled the farm's development were either unaware or unsupportive of such alternative economic structures.

For still other locals, Blueprint Farm was simply a "Jewish thing," unrelated to the pattern and conditions of their lives. Even local sophisticates, like the architect Rafael Longoria, found the appearance of the farm to be alien— "more like El Paso than Laredo."[12] The reception by active farmers, however, varied. Ironically it was the midsized commercial growers—those who were already doing well—who demonstrated the most interest in the Israeli technology. The smaller, marginal growers—those the farm was intended to benefit—were mystified by the capital-intensive nature of the project, especially the Israeli drip irrigation technology. They perceived the farm as simply another government research subsidy to those who don't need it. Like the small farmers, the faculty of Laredo Junior College was generally uninterested in the farm. Lack of interest may have been, however, the feigned attitude of those who felt slighted by the short-lived attention granted the newcomers by the college administration. One biology professor did become involved in the project when it was nearly over, but his students were something less than enthusiastic about taking lessons from either the ecologists or the Israelis. The experimental produce grown by them for course credit rotted in the field as soon as classes were over. In general the reception by local farmers and citizens not invested in the farm's production was one of indifference. Where local experts were anxious to receive notoriety and praise through their global networks, local small farmers were anxious to avoid risk.

Miscellaneous support did, however, materialize in the community. A few politicians, the news media, the Centro Aztlán Food Bank, and others were receptive to the idea of alternative technologies that might advance their own interests. But for each unsolicited gift of goodwill, there was ample poison to go around. The college administration, in addition to the Israelis, did not receive the ecologists' program of organic production with anything but derision. The only locals who really supported the ecologists were the workers hired by Fisk to actually build the structures. Several of these workers became converts to Fisk's organic vision. For example, the construction crew chief, Thomas Rosas, became convinced that there was, as he phrased it, "a direct relation between the health of the soil and the health of people—to rob one was to rob the other."[13] He remains convinced, as was Karl Marx himself, that agribusiness is the robber of "the source of all wealth—the soil and the labourer."[14]

Believers are always beset by nonbelievers. As their position became more and more peripheral, the ecologists began to see conspiracies against their interests everywhere. After the closure of the farm, they resigned themselves to the bitter (and self-serving) view voiced by Alvaro Lacayo that "the organic vision can't happen until the mechanical technology falls of its own weight."[15] What little support and recognition the ecologists did receive came principally from the outside—from regional or national sources. It was this link to global networks, and the geographic isolation of the project, that contributed to the emerging cult status of Blueprint Farm within the environmental movement and on the margins of the architecture profession. The ecologists were shocked to discover that the Israelis had more in common with the land grant network than with themselves. To find themselves coupled for political purposes with the techniques of maximized production at best was confusing and at worst felt downright exploitative. They received the unfolding situation with dystopian gloom.

Most architects, even local ones, learned about the project in the May 1991 issue of the journal *Architecture*—one of the most influential national professional journals. This level of legitimization in the national press opened up a new discourse to Fisk and the ecologists. In the journal's review of the project's structures, the formal references to the work of the Philadelphia architect Louis Kahn, and the technological references to the inventions of the Texan architect O'Neill Ford, were immediately recognized by professional readers. Some readers, like the architect Rafael Longoria, admired the "minimalist phenomenology" of these structures. For these admirers, the simple orthogonal geometry lent the structures a formal quality that commanded respect. Other architects, however, found the rawness of the materials and detailing to be romantic and something less than "architecture." For these detractors, the heaviness, or earthiness, of the structures precluded the "spiritual transformation that is required by architecture." Drawing specifically from Kenneth Frampton's discourse on the "stereotomic" (the low-tech heaviness that aspires to the earth) and the "tectonic" (the high-tech, light constructions that aspire to the heavens), the architect Rafael Bernadini lamented that "the juxtapositioning of the heavy and the light didn't happen" at Blueprint Farm.[16] There was, for Bernadini, only the heavy dumbness of rural life at Blueprint Farm—a condition from which he worked hard to distance himself. He reflected that the ruins of the farm felt like "Perth Amboy ... powerful, but deserted."[17] Architects like Bernadini clearly understood the regionalist attitude deployed at Blueprint Farm, but only in formal terms. And with equal clarity professionals rejected the "counterculture technology" or "the barn-raising culture" employed in the farm's construction "as appropriate means for architecture." Behind these critical views is the demand that whatever is "good" in a building must be "readable." In other words, architecture for these Texas professionals is received as a visual and formal construction, accessible only through educated eyes. Because the Israeli contributions to the landscape were either underground or vegetative, they didn't exist for the architects. These technologies weren't received at all because to professional eyes they were invisible. In this sense, architects tended to receive the farm with the same kind of categorical blinders as did the TDA bureaucrats who introduced this story in Chapter 1. In their common view, agriculture and architecture are unrelated practices.

The land grant network received the farm on terms entirely different from those of the architects. Pat Roberts, the same senior AES official who has been previously quoted, has best summarized the attitude of the land grant network toward Blueprint Farm as the reification of Jim Hightower's agriculture policy. Roberts complained that "the myth of direct marketing, and the purpose of the farm, was to turn farming back to forty acres and a mule." With incredulity, Roberts argued that "you don't feed 17 million Texans by selling fruits and veggies from the back of a pickup truck." The agricultural methods promoted at Blueprint Farm, which Roberts derided as "low-tech," won't provide for the good things in life—"for pizza on Friday night, a little TV, and some schooling."[18] The loss of productivity that would result from Hightower's policies would, in the eyes of Roberts and his colleagues in the land grant network, devastate the material standard of living enjoyed by Americans—including themselves.[19] To say that the land grant network, and their constituents, were threatened by, and hostile toward, Blueprint Farm is an understatement of heroic proportions. That such a threat to their security could be mounted by Hightower using a technology "invented" by AES (the so-called "Israeli" drip irrigation system) was an irony so bitter that AES officials could barely discuss it. Not unlike the ecologists, the land grant network began to see a conspiracy in the threat to their material well-being. For Pat Roberts, the conspiracy could only be the work of Others—Jews, hippies, and homosexuals.[20] These aliens spoke a language unintelligible to, as Roberts put it, "normal folks." As a result, they were not entitled to treatment according to the same ethical standards of behavior as those who "speak our language."[21] Roberts personalized the threat that he perceived to be embodied in Blueprint

Farm by anxiously reminiscing about the trajectory of his own life: "I was born on a tobacco farm and I sure don't want to go back to 'Tobacco Road.'" He projected the same deep fear of deprivation on his wife's account: "When she starts my dinner, my wife wants to open a can of carrots, she doesn't want to start with *raw* food."[22] As remarkable as such an allergic response to "rawness" seems in our current cultural situation, the pathos in Roberts's interpretation of reality is striking. It seems that once a young man leaves the farm, he—like the migrant farmworkers of Laredo—would endure almost anything to avoid returning to what he perceives (ironically, like Karl Marx) to be the "idiocy" of rural life. The land grant network received the farm as a threat to American productivity and to their positions of privilege in the political apparatus of technoscience. The enemy was received without ambiguity, but with a great deal of anxiety.

If the land grant network received the farm with open hostility, the market received its products with skepticism. Of course, the financial naïveté of both the Israeli and American farm managers didn't help. Although productive technology was in place, and valuable niche crops had been identified, a distribution mechanism, or market strategy, for such exotic fare as Persian melons and bell peppers was never developed by TDA as promised. The conventional assumption behind a market strategy is, of course, that no farm could survive without being integrated into the existing system of production and distribution dominated by the land grant network. To his credit, Hightower had the courage to imagine that an alternative, or independent, system of production and distribution might be developed. Tragically, however, neither he, his supporters at TDA, local experts, nor the local managers of Blueprint Farm had the courage to actually construct such a noncapitalist system of support. As a result, commercial and noncommercial buyers alike remained distant while crops rotted in the beautiful, but un-air-conditioned, packing sheds.

Not only did buyers remain economically distant from the farm, but also the public remained physically distant. Not for ideological reasons, nor for lack of information, but because the site of Blueprint Farm, cut off from town by the railroad and a bend in the river, was invisible to them. The physical seclusion of the farm contributed to its local identity as a "science" place, or an "art" place, isolated physically and psychologically from the daily lives of locals. Henri Lefebvre argues that art has degraded everyday life by insisting that the true and the beautiful arise only from the "marvelous." Martin Heidegger has likewise argued that science has degraded everyday life by insisting that the true arises only from the sublime. Once reality is so transformed, the region of art is alienated from the conditions of everyday life and is controlled by priests, magicians, or poets. Similarly, the region of material life is alienated from the conditions of the everyday when it is transformed into technoscience and controlled by scientists.[23] That Texans like the ecologists were denied authorship of whatever art or science they contributed to the project (in the name of Israeli hegemony) only magnified the local impression that the farm was born by immaculate or foreign conception. Of course, efforts were made to import witnesses inside the barriers created by local geography. But the small farmers and schoolchildren who were treated to tours of big tomatoes were themselves invisible citizens. Their testimony didn't count for much in the courts of politics or science—a lot of invisible people are still invisible when they are politically silent.

An invisible place belongs to no one. The farm had no owner, or no potential owner was permitted to inhabit the place by the college bureaucracy. Although all of the networks wanted a piece of the action, nobody (save the ecologists) even talked about living on the farm. As the ever-sharp Alvaro Lacayo put it, "Everybody wanted to seize the limelight—they were naïfs who enjoyed the ride." In the void of ownerlessness, bureaucratic order was imposed. In response to the daily crises of farm production, the college, rather than expedite solutions, mandated what Lacayo described as "a three-day purchase order process to buy a ten-dollar part." The LJC trustees were willing to tolerate the farm if it "put them on the map and made history." But they were not willing to tolerate paying for that history-making opportunity. Lacayo wryly noted in 1995 that "the college would now give the farm to anybody who offered to cut the grass."[24] The college welcomed the farm only as an opportunity unburdened by responsibility.

If the college trustees were accused of opportunism by some, others accused individual Israelis of the same sin. The first Israeli manager was lured away to a large commercial farm down the river by the prospect of profit. The second Israeli manager was more intent upon assessing market potential for equipment sales than on farming or teaching local farmers to actually use the imported hardware. When the market for Israeli products never materialized, the second Israeli manager departed as well. Even the Texas-born manager hired to replace Israeli management was lured away by commercial interests once he learned how the irrigation system worked. Nobody stayed. The accusation of opportunism extended to the ecologists, to Hightower, even to the farmworkers themselves. Ecologists, including the Texas Organic Gardeners Association, found the international exposure they enjoyed on the farm to be a heady experience. "It was a trip," giggled Alvaro Lacayo. Of course, Hightower's intentions were routinely received with suspicion. And when the farm was finally taken over by local activists after the 1990 election, workers at the farm were paid well, but were not required to actually *do* anything. The farm became a purely instrumental place for those with distant aspirations—a place of reference, not a place to live.

The multidimensional competition among those seeking opportunities—places to maximize their own interests—was masked by the two-dimensional confrontation between the ecologists (principally Pliny Fisk) and the Israelis (principally Benni Gamliel, the second farm manager). The creation of the demilitarized zone that territorialized these two sets of interests had the effect of concealing the frustrated interests of other networks. Although attempts at mediation were made, the duality of "high-tech vs. low-tech," or the Israelis vs. the Texan ecologists, forced people to take sides. Ultimately, the scales of proof, at least in the eyes of locals, tipped on the side of the Israelis. Although the forces of the "high-tech" won the campaign, both sides lost the war.

As Robert Holub has suggested, each of the networks investigated here understood Blueprint Farm from inside its own interpretive paradigm. But if the reader will recall, Holub argued that "a given paradigm creates both the technique for interpretation and the objects to be interpreted." In other words, we should understand that the objects constructed at the farm, as well as the modes of interpreting them, were produced by conflicting assumptions about the nature of reality. Figure 6.1 will help to clarify the relationship between interpretive paradigms and the objects created at Blueprint Farm by relating the concept of intentionality discussed in Chapter 4 to Jauss's concept of reception. In making this leap from the interpretive structure conceived by Husserl and Heidegger to that conceived by Jauss, I am not suggesting that these three philosophers share common assumptions—clearly there is too much uncommon epistemology between them to permit easy hybridizing. For the purposes of this study, however, it is reasonable to relate pragmatically the structure of intentions to the structure of receptions.

In this construction I am suggesting that three of the seven networks identified—the Israelis, the land grant network, and local experts—received Blueprint Farm through a common interpretive paradigm that is both *market-driven* and *high-tech*. By market-driven I simply mean that Israelis, agribusiness executives, and local business leaders understood their work to be the production of exchange-value, not the production of artifacts with use-value. By high-tech, I mean that the members of these networks shared a predisposition toward technological determinism and the promise of a high-tech fix to their problems.

Unlike the cohesive pastoral community of Las Trampas in New Mexico, the small farmers of la Frontera Chica are neither a homogeneous group nor peasants with an archaic, organic connection to the land. These marginal producers identify with the economic aspirations of the Israelis, the land grant network, and local experts, and they explicitly accept the conditions of the market as relevant to their lives. The difference is, however, that where local

THE INTERPRETIVE PARADIGM	OF ANY NETWORK		IS SATISFIED BY	THE PRODUCTION OF THINGS TO INTERPRET
The market-driven,	high-tech	Israeli network	is satisfied by	profits.
The market-driven,	high-tech	land grant network	is satisfied by	tons of goods.
The market-driven,	high-tech	local expert network	is satisfied by	black boxes.
The market-driven,	pragmatic	local farmers' network	is satisfied by	useful tools.
The formalist,	instrumental	local Jewish network	is satisfied by	icons of goodwill.
The formalist,	instrumental	architects' network	is satisfied by	beautiful objects.
The	low-tech	ecologist network	is satisfied by	sustainable systems.

Figure 6.1. Conflicting interpretive paradigms operating at Blueprint Farm.

experts and their collaborators expect the market to serve their interests, the small farmers of la Frontera Chica expect uncertainty and exploitation from the market. Influenced by similar logic, their attitudes toward technology are dramatically more skeptical than those of local experts who identify with market forces. Such dismal expectations are rationally derived from years of experience. Small farmers, then, are invisible and reluctant participants in an interpretive paradigm constructed by others. This condition is, critical theorists would argue, an operative definition of *hegemony*—a form of domination and exploitation so pervasive that even those most affected by injustice perceive it to be neutral and natural. Local farm activists, however, Alvaro Lacayo principal among them, suffered from no such false consciousness.

Two of the interpretive groups—local Jews and Texas architects—shared an interpretive paradigm that is *formalist* and *instrumental* with regard to technology. By formalist, I mean that the members of these networks interpreted meaning through reductive visual strategies. By instrumental, I mean that the members of these networks assumed that technological means need not be related to the aesthetic ends that they value.

This leaves the ecologists isolated, like local farmers, but in an interpre-

RECEPTION 149

tive paradigm of their own making. I'll characterize their mode of interpretation as *noncapitalist* and *low-tech*. By noncapitalist, I mean that the ecologists understood reality to be the production of use-value to the community, not an abstract exchange-value. By low-tech, I mean only that the ecologists were concerned with the ontological relationship between technological means and natural processes, not that they favored primitive or nostalgic modes of production. In this sense the Israelis and local experts were correct in receiving the ecologists as alchemists rather than scientists. As Arnold Pacey would argue, the "participatory attitude" of the ecologists left them isolated in a field of pseudoscience.[25]

However, the meta-categories that I have constructed here—the market-driven, the formal, and the noncapitalist—present two problems. First, the interests of the formalists were rendered nearly invisible by the battle for imaginative supremacy that was waged between the marketeers and the ecologists. As a result, a formal interpretation of the project has not been seriously considered. Second, the ecologists did not in any way receive their own intentions to be "low-tech." They would object, no doubt, to this characterization of their intentions. It is necessary to briefly examine these two seeming inconsistencies.

Those whom I refer to as formalists—local Jews and architects—were only indirectly involved, if at all, in the production of Blueprint Farm. For the most part, they simply received and interpreted the works of others. It is not surprising, then, that their way of interpreting the world through appearances, stripped of ecological and political context, had little influence in the turf struggle between marketeers and ecologists. It seems reasonable to argue, however, that the modes of interpretation employed by both the formalists and the marketeers shared a similar reductive strategy. Limiting the meaning of a thing to relationships derived from its formal, compositional characteristics, or to its exchange-value, is at least *methodologically* similar. In the world of high art, the compatibility of the reductive interpretations made by marketeers and formalists has often created common cause and common space (i.e., museums). However, in this case, *appearances* and *profits* were not perceived as commensurable; thus the marketeers and the formalists found themselves to be at odds with one another. By interpreting Blueprint Farm in instrumental terms, the formalists did offer a way out of the opposition between the high-tech and the low-tech perceived by marketeers. For marketeers, however, the formalist interpretive paradigm seemed irrelevant to the economic conditions of daily life along the border. As a result, the formalist interpretation of Blueprint Farm was suppressed. Had marketeers and formalists discovered their common reductive interpretive methodology, things might have turned out differently.

The second problem made visible by my categorizing of interpretive paradigms in Figure 6.1 is identifying the ecologists as "low-tech"—being placed in such a box would simply baffle them. They saw their project as combining the local and the universal, the high-tech and the low-tech, the tectonic and the stereotomic, in much the same way that the formalists (and Frampton) have proposed. In spite of their intentions, however, the Israelis, the land grant network, and the architects in particular *received* the technological objects constructed by the ecologists as earthbound and low-tech. Local farmers, however, categorized the schemes of the ecologists as alternately too high-tech, or too low-tech, but never just right for their needs. I would suggest that we understand this paradox by employing Robert Holub's insight regarding the importation of concepts from one culture to another. If the reader will recall, Holub concluded that importation could only be successfully accomplished if the receiving culture has an "already established constellation of ideas" to which the import might be grafted. In Chapter 3, "The Local History of Space," I portrayed la Frontera Chica as having no available tradition of ecologism or progressive agrarianism to which the ecologists might have attached their machines. Fisk and his collaborators were inventing into a landscape seemingly void of kindred ideas. The concept of sustainability was dramatically more alien to local farmers than were the market-driven development schemes of the Israelis. I want to stress that to say that the market economy was *familiar* to local farmers does not mean that they perceived it to act in their interests. To the contrary. It is just that the concept of profit and international trade had become, or better, has always been, the local mantra of those who hoped for change. The concept of sustainability, in contrast, was a foreign language. As a result, it is hardly surprising that the machines of the ecologists were received and interpreted otherwise than intended.

Long after the disappearance of the Blueprint Farm it is easy to lament that its builders failed to relate the concept of sustainability to those local cultural traditions that now seem to be its natural allies. The borderland concept of *mestiza* (where space to experiment resides in the interstices between cultures), the concept of *ejido* (the communal landholding practices of New Mexico), and the concept of *vergüenza* in relation to the communitarian ideal of cooperation are all indigenous concepts upon which the imported notion of sustainability might have been built. It is reasonable to argue that such a critical perspective might fail to recognize just how undefined the concept of sustainability was in 1987. That same perspective, however, might prevent future builders from committing comparable acts of cultural insensitivity.

In the struggle for imaginative supremacy, the Israelis found local ground to be more fertile than did the ecologists. The Israelis eventually discovered, however, that transplanting their imported economic and political apparatus was impossible. This observation is supported by the interpretation of technological objects proposed in Chapter 5. If the reader can recall, I argued there that technological objects contain political properties that are culturally specific and related to the place of their production. As a result it is hardly surprising that the machines of the Israeli kibbutzim, like those of the ecologists, were received and interpreted otherwise than intended.

It would, of course, be futile to blame the ecologists for failing to relate their project to local "constellations of ideas," or, as Latour would put it, failing to mobilize resources that would support their interpretation of reality. In practice the ecologists planned several "outreach programs" which were designed to introduce the concept of sustainability into the everyday lives of locals. Community gardening and composting, a food bank, continuing agricultural education, and worker housing were all proposed by the ecologists and rejected by local experts and TDA as nonarchitectural activities. Without the administrative support of Laredo Junior College or the economic support of TDA, however, the intentions of the ecologists can only be understood in terms of their own interpretive, or ideological, isolation. To blame the ecologists alone for failing to mobilize the support of all the actors in this story would require us to accept the great-man theory of history that was rejected in Chapter 5.

Not to blame the ecologists, however, is not to excuse them. Like the other actors in this story they assumed that local farmworkers had not developed a critical consciousness of sufficient power to change conditions for themselves. Likewise, these intellectuals assumed that local farmworkers had nothing to say that might inform the project they conceived. In the best of social activist traditions, however, the function of the intellectual in society is to help locals understand the systematic character of local power imbalances. Once local workers are equipped with a global understanding of their situation they are perceived as capable of producing change on their own behalf. On this view, the revolutionary role of the intellectual is *dialogic,* not prescriptive. Thomas McLaughlin puts it this way:

> The true revolutionary engages in dialogue with the oppressed, trusting their insights, not knowing in advance how the process will turn out, not expecting an end to the process of coming to consciousness.[26]

In the case of Blueprint Farm, however, the "revolutionaries" arrived in Laredo knowing in advance exactly what they wanted to build.

The interpretive tragedy of Blueprint Farm was that it was received, not as a creative dialogue between the forces of high-tech and the low-tech, or between the stereotomic and the tectonic (as Frampton would see it), but rather, by local farmers in particular, as an antinomy—an unresolvable contradiction—between high and low technology. Because the producers of the farm did not understand, nor seek to discover, the interpretive paradigm of those small farmers expected to receive it into their lives, those farmers perceived themselves to be the reluctant objects of production, not the subjects of conception. Under such undemocratic conditions no technology transfer could take place.

The formalists did, however, offer an instrumental resolution to the conflict by ignoring the opposition between the technological means employed by both marketeers and ecologists. For the formalists, the opposition between the organic and the chemical, or between the low-tech and the high-tech, was lost in the subjective construction of poetic meanings. But in this case, no such aesthetic distraction from material conditions was allowed to stand. The aesthetic programs of local Jews and Texas architects were simply ignored by those in control of the site. The experience of Blueprint Farm was the symmetrical negation of an economic interpretation of place and an ecological interpretation of place. The perceived opposition between high and low technology, rather than informing the history of Blueprint Farm, as critical theorists would have it, precluded its realization as an inhabited place. This interpretive standoff requires that we next consider the question of *reproduction* as the subject of Chapter 7. It is through this concept that the limits of critical theory as an interpretive paradigm, and of critical regionalism as a polemic of production, can be made clear and thus prepare the ground for a final proposal in Chapter 8.

CHAPTER SEVEN

REPRODUCTION

If space is a product, our knowledge of it must be expressed to reproduce and expand the process of production. The "object" of interest must be expected to shift from *things in space* to the actual *production of space.*

Henri Lefebvre, *The Production of Space,* p. 37

The problem of the builder of "fact" is the same as that of the builder of "objects": how to convince others, how to control their behavior, how to gather sufficient resources in one place, how to have the claim or object spread out in time and space.

Bruno Latour, *Science in Action,* p. 131

In these two passages, Lefebvre and Latour define the problem of *reproducing* one's intentions. The purpose of this chapter is to investigate how, or if, the production of Blueprint Farm has established technological facts that have "spread out in time and space." The structure of this chapter will by now seem familiar: First, I will review the historical methods of "fact production." Second, I will reconstruct the goings-on at the farm before, third, subjecting those goings-on to an interpretation. I find that two Israeli technologies were successfully transferred to the private sector elsewhere in Texas. However, the utopian project of the ecologists was reproduced primarily in a nonmaterial, sublime context. In the conclusion of this chapter, David Nye's concept of the *technological sublime* proves to be helpful in understanding the problem of reproduction that was encountered by the ecologists.

In common usage, *reproduction* is a biological metaphor for the concept of multiplication. In the biblical imagination, as in the imagination of marketeers, to reproduce is synonymous with the dual concepts of *prosperity* and *domination*. The faithful reproduce by conquering nonbelievers. Likewise, capital reproduces by conquering markets. As do humans, rabbits, or fire ants, technological networks conquer spaces through the act of reproduction, or multiplication. In Latour's terms, a technological network dominates space by incorporating potential detractors, and their artifacts, into its ranks. Each conversion is in fact a twofold multiplication—the loss of one potential detractor and the gain of one supporter. The reproduction of facts, believers, or consumers becomes, like population growth, an algebraic calculation.

To further illuminate the concept of reproduction, it might be helpful to distinguish it from concepts more typically associated with architecture: the emerging concept of the *sustainable* and the traditional concept of the *durable*. In Chapters 1 and 3 we briefly considered the developing professional definition of sustainability. In everyday parlance, however, to sustain is simply to keep in existence, to maintain balance, or to provide for. As John Lyle has lamented, to sustain is to reproduce without the requisite of expansion or improvement to already degraded conditions. Compare this to the common definition of the durable—which means only to last for a long time, or withstand wear and tear. Where the concept of reproduction assumes the necessity of growth in order to survive, the sustainable assumes only the necessity of maintaining the status quo, and the durable wishes only to delay the inevitability of decay. What most distinguishes the concept of reproduction, then, is the implicit mandate to produce new spatial conditions through expansion. Each of the technological networks engaged in the production of the farm expected their intentions to be reproduced as facts—to dominate more and more space over time. In Lefebvre's view, one cannot "change life ... without the production of an appropriate space."[1]

Geographers tend to argue for a model of technological diffusion that relies upon a core/periphery dynamic, in which technologies migrate from urban centers, then to neighborhoods, and finally to the outlying hinterlands.

THE PRODUCTION OF FACTS

Jennifer Tann, however, finds this model to be highly "stylized," in that it assumes geometry, rather than technological networks, to control the process of diffusion.[2] This section provides an alternative way to understand the reproduction of technologies that is based upon the social production of facts and their diffusion through spatial networks.

From their study of the origins of Western science in the seventeenth century, the sociologists Steven Shapin and Simon Schaeffer have concluded that modern science has historically relied upon three methods to first produce, and subsequently reproduce, scientific facts. These are the *material,* the *literary,* and the *social.*[3] Before describing how the methods of fact reproduction operate, however, I want to reinforce Latour's claim (cited above) that there is no effective difference between how scientists use such methods in the construction of facts and how architects use them in the construction of objects. Scientists and architects are, in this view, both builders. Both have to "convince others," "control behavior," and "gather resources" in order to make their claims spread. The insights of Shapin and Schaeffer are helpful in demonstrating how Blueprint Farm, in spite of its appropriation and conversion for other purposes, has been reproduced as an architectural "fact" within the canon of the emerging ecologist network.

The material method of fact production, as defined by Shapin and Schaeffer, is achieved by demonstrating the working apparatus of the experiment to potential supporters. In contemporary language, Americans would refer to material technology as *hardware.* This category refers to the "sets of objects" described by MacKenzie and Wajcman in Chapter 5. For architects, of course, the working apparatus of the experiment is the functioning building itself. Thus the working apparatus, or the building, is the principal evidence that scientists or architects produce in order to win supporters. It is difficult practically, however, to have legions of potential supporters traipse through your laboratory or, in many cases, through your client's building.

The literary method of fact production helps to suppress the messiness of materiality. It is achieved through the testimony of those who were not firsthand witnesses to the (scientific or architectural) construction, but were conscripted into being supporters of the claim. According to Shapin and Schaeffer, seventeenth-century science was socially constructed as a metaphor of "public space." Scientific events, as opposed to the private, mysterious events of alchemy, were "witnessed" in a legal sense by an established academy of peers. It is the social act of witnessing—in the laboratory or classroom, or at scholarly conferences—that is the essence of modern fact production. Facts, then, are matters of rational social agreement between authors and witnesses. The scholarly and popular publications that examine science and architecture provide the means to construct an abstract space of "virtual witnessing" in the reader's mind. These publications are narratives of sensory experience. Because they rely heavily upon visual images, the reader experiences a controlled simulation of reality. What is produced by the literary method of fact production is an "experimental community," or a community of professional interest, that extends the public space of the laboratory, or the building site, to those who are potentially affected by the claim.

The social method of fact production is, like the literary, concerned with networks of testimony. It is achieved through the development of social conventions and networks upon which potential supporters might rely in evaluating or reproducing knowledge-claims. The scientific or professional conference is a forum that tests the social viability of truth-conventions. Through the performance of argumentation, supporters to a claim are mobilized by demonstrations of how their interests are contained in the fact being promoted. Universities and architectural firms are similarly part of the social method of fact production. Through the system of tutelage, students and interns are rewarded financially and socially for witnessing and reproducing the facts claimed by their mentor.

In Latour's terms, the act of witnessing the production and reproduction of facts is the basis for the construction of both the facts themselves and the technoscientific networks that depend upon them. Participation in the social act of fact production, by subscribing to journals, attending conferences, or being a student, is tantamount to reproducing the author's claim.

By referring to the Appendix, which briefly describes the **SPREADING** remaining nontraditional technologies incorporated at **CLAIMS** Blueprint Farm, the reader can count which of these have been reproduced by others. On the Israeli side of the ledger, large growers have successfully reproduced two of the imported technologies in Texas—Fertigation™ and "inorganic mulch." On the ecologists' side, however, the twelve Biom-metric™ technologies have been largely ignored, been appropriated for other uses, or have lapsed into ruin. A particularly ironic example is that the wind generators and pumps, originally designed to irrigate experimental crops, have been used (when they work) to water the college lawns. CMPBS is quick to point out, however, that the dysfunction of the project's water system was the responsibility of Burgey Wind Turbines, Inc., the manufacturer of the wind turbines.

Many of the project participants have wondered out loud about the apparent success of experiments with alternative technologies in Israel and the apparent failure of experimentation in America. It doesn't take a great deal of cross-cultural study to realize, however, that in contrast to conditions in Israel, there were serious economic disincentives operating in Texas that contributed to the low rate of reproduction of all Blueprint Farm's nontraditional technologies. Pat Roberts, the previously cited AES official, argued that food production is, in Israel, a matter of "state security." Not only do the kibbutzniks produce food, but by occupying contested land they operate an early warning system against Arab invasion. Under such conditions, the money economy is a relative consideration. A second rationale for the poor American showing can be found in Alvaro Lacayo's observations about the difference between political conditions in Israel and la Frontera Chica. The Israeli government, he holds, is both committed to the social experiment of kibbutzim and to the elimination of hunger in the general population. That commitment is in distinct contrast to the market-driven agriculture policy propagated in the United States. Local farmers familiar with Blueprint Farm also understood that, where Israeli agricultural policy supports small-scale production and local consumption of produce, U.S. agricultural policy supports large-scale production and global consumption of bulk grains via the international commodities market. In the disparity between these modes of production, farmers in la Frontera Chica felt torn. Although some were interested in the new Israeli techniques, without a dramatic shift in government agricultural policy the economic risks were simply too great to take the chance. With self-deprecating humor, the farmer Roberto Elizondo claimed that "economics and ignorance is what makes us tick."[4] The disincentives for investment in alternative technologies were overwhelming and the reasons for adopting these technologies were opaque to those who stood to lose the most.

Had local farmers witnessed an expanding market for the farm's exotic produce, they might have been encouraged to take a few risks. The reverse, however, was the case. Alvaro Lacayo noted that "nobody [at Blueprint Farm] planned. Stuff rotted in the fields."[5] Eventually, even Dr. Jiménez—the committed technological determinist—recognized that increased productivity is useless unless "a market strategy is already in place."[6] Inside this repentant insight was the recognition that TDA, and Jim Hightower, had failed to deliver the promised market support system. The result was a perpetual state of crisis.

Not unlike many marginal farms, Blueprint Demonstration Farm operated on the economic edge. Promised funds arrived either late or not at all. In addition to experiencing economic insecurity, the managers of the farm were un-der the constant pressure from TDA to demonstrate this or that crop to Hightower's potential political supporters. In such a state of perpetual experi-ment, no continuity, or economy of scale, could develop. The mandate to dem-onstrate exotica, as if the farm were a sideshow at the county fair, was a serious diseconomy. The burden of agricultural demonstration, however, did not re-lieve the farm of the political mandate to demonstrate economic "self-suffi-ciency." As the rhetoric of the project gained momentum, most observers conflated the concept of "self-sufficiency" with that of "sustainability." In the face of such confused expectations, nearly all participants argued that the farm could have demonstrated economic viability had funding been continued long enough to complete the "fine-tuning" of technological systems and establish a market toehold. Most agreed, however, that in the conservative political climate of the 1990's, such public support was unlikely. One unanimous agreement did emerge: that what the farm needed most was a "farmer"—an independent yeo-man molded in the democratic model constructed by Jefferson.

The final disappointment of local experts was that the "scientists" failed to produce what Dr. Juárez described as a "knowledge breakthrough."[7] Initially, project participants and local farmers had welcomed "as signs of success" what Alvaro Lacayo rattled off as a litany of "ribbon cuttings,""impressive structures," and "hoopla." Eventually, however, when the thrill faded and the Israelis went home, local farmers derisively characterized the farm as "a big political deal," and "a showplace for dignitaries." Only after the fact did local farmers figure out that "people didn't *learn* anything. They just saw big tomatoes."[8] With hindsight, even the college trustees came to the realization that the Israelis had never engaged in teaching or even in writing the training manual they were con-tracted to produce. In lieu of the cross-cultural exchange so dramatized by lo-cal Jews, or the technological fix imagined by local experts, what Laredoans received was a demonstration of sublime technological superiority by Israelis and a demonstration of quixotic utopianism by the ecologists.

But as uninterested in teaching as the Israelis proved to be, local citizens proved equally uninterested in learning. LJC students barely walked onto the site. With similar aversion, the local Job Corps Center rejected involvement in the farm because their trainees had absolutely no interest in agriculture. Although some responded favorably to the "high-tech image" of the place, as soon as that image was contaminated by the social history of what farmworkers call "stoop work," it tarnished quickly.[9] In the absence of teaching and learning, it was not difficult for Jim Hightower's successor, Rick Perry, to, as Pat Roberts put it, "reestablish the *facts* of Texas agriculture."[10] Without local reproduction of the considerable knowledge generated at Blueprint Farm, those who had an interest in suppressing change had a rather easy time of writing the history of the farm in their own terms. That official history, of course, maintains that Blueprint Farm failed on *economic* grounds.

Given the mammoth annual state and federal subsidies provided to the land grant network, the accusation that this tiny experiment failed to stand on its own two feet is particularly ironic. The accusation is, of course, an example of economic determinism selectively applied. Those who argued most vehemently against the meager support granted to Blueprint Farm—Pat Roberts, for example—were those who received major support from public sources. Rather than recognize the differing kinds of value provided to society by agrarian practices, as is apparently the case in Israel (as well as England and France, just to name a few examples), U.S. agricultural policy is increasingly based upon the reductive concept of economic value. In the language of sustainability, such narrow economic determinism implicitly devalues the ecological and social dimensions of that concept. Social ecologists argue that all three dimensions of sustainability discussed in Chapter 4—the economic, the ecological, *and* the social—are practically required to achieve reproduction.

That so many of the participants in this project rationalized its disappearance on economic grounds suggests that even those who were ideologically opposed to the dominance of the land grant network inadvertently portrayed its authority as inescapable. Just as science was portrayed by Dr. Jiménez to be objective and neutral, capitalism was portrayed by the project's builders as a "singular, unified, and total system."[11] The tragedy in this portrayal is that it perpetuates the very conditions that the builders wanted to escape. By passively blaming Hightower and TDA for not delivering the promised market strategy, the builders of the farm subjected themselves to the authority of the market as it now exists. The difficult alternative, of course, would have been to produce and reproduce a *mestiza* system of distribution—one that would exist within the interstices of dominant technological networks.

Not long before the 1990 election, Hightower was so frustrated with the indifference of Laredo Junior College toward the farm's political mission to construct such an alternative network that he transferred administrative responsibility to Texas A&I University. His strategic plan included the seemingly subversive intention of integrating the farm with the land grant network itself. In South Texas, Texas A&I is the representative of the land grant network's interests. It would seem, then, that if he had failed to construct an alternative, or *mestiza,* technological network, Hightower intended to reform the land grant network from within. What Hightower did not foresee, however, was that this last lifeline would exploit grant sources for its own benefit even more voraciously than did LJC.

When Hightower lost the election a few months later, the loss of his sponsorship provided a convenient rationale for the closure of the farm. Astute observers like Alvaro Lacayo recognized that "the farm was born politically and died politically."[12] It was certainly suppressed from the outside by the Perry administration, but it also failed to provide internal rationales for its own reproduction. In the eyes of the larger agricultural community in Texas, Blueprint Farm had become so linked to the personal fortunes of Jim Hightower that it was a logical icon of retribution for those who perceived it to be a launching site for his political ambitions.

When Hightower initially took office in 1980, those career bureaucrats who survived the purge of personnel loyal to the previous administration went underground. Ten years later, when Hightower was in turn purged by voters, those same bureaucrats who went underground in 1980 reemerged from their bunkers to inherit the spoils accumulated by the regime of sustainability. It was these individuals who proved more *durable* than the technological objects planted by Hightower and the ecologists. After ten years in hiding, affiliates of the land grant network had, however, learned a valuable lesson. If they had no intention of reproducing Hightower's politics, they had learned the value of his rhetoric. The project of AES became the appropriation and redescription of what "sustainability" might mean in the post-Hightower era.

Late in 1989, Jim Hightower did hold "secret talks" with Texas A&M officials about establishing "an institutional base for Blueprint Farm." Hightower understood that such an institutional base would be required for the Blueprint Farm technologies to be successfully transferred into the industry. Like many other interested observers, he had come to the conclusion that, without the constructive involvement of the land grant network, the farm and its mission would remain in the margin of agricultural practice.[13] But by this time it was too late—land grant officials at Texas A&M smelled blood in the coming election. As a result, Pat Roberts claims, the "technology transfer never happened."[14] Texas A&M officials simply waited to witness the closure of the farm and then picked up the pieces.

While Hightower held his secret liaisons with the internal enemy, other TDA administrators courted a series of potential external supporters. After the Israelis departed in 1990, a group of Chinese investors supported a one-season experiment in the production of organic vegetables bound for the Houston-based Asian market. The Russians came to visit and, for a time (much to the dismay of the FBI), demonstrated serious interest in the biological enzymes used by the ecologists as fertilizer. Also proposed were ties to Mexican agricultural universities with established links to Israel. But it was only the ecologists who argued that it was local small farmers who were the "missing link" to reproduction. The exhortations of Alvaro Lacayo to engage farmworkers directly in the farm's management structure were, however, lost in the din of the last few months of desperate grant writing.

After Blueprint Farm sat deserted for a few years, the Rio Grande Interna-

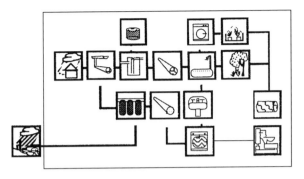

Figure 7.1. Icono-metric™ diagram of water systems, by Pliny Fisk III, © 1994 CMPBS. Courtesy of the Center for Maximum Potential Building Systems.

tional Study Center (RISC), in exchange for conducting minimal maintenance, like cutting the grass, was given the use of the property by the college. RISC is a nonprofit river-study, river-watch organization with a pedagogical mission that centers on the water resources of the Rio Grande that are shared by Mexico and the United States. The founders of RISC are confident that "the site has more potential to involve the community as a river-study center than it ever did as a farm."[15] The Meadows Foundation, the principal benefactor of Blueprint Farm, apparently agreed because it has funded RISC generously.[16]

Faced with the political and material demise of Blueprint Farm, Pliny Fisk continued work on his *Biom-metric*™ and *Icono-metric*™ technologies as tools of reproduction. If the farm could not be salvaged in *real* space, he reasoned, his version of reality could be reproduced in hyperspace. Icono-metric™ representations are computerized architectural narratives that empower the user to tell a story about a place by engaging design decisions with the interactive natural systems present at the site. In the most positive sense, this computer program assists designers in unpacking the black box of scientific facts and associated interests constructed by the land grant network. The software hopes to accomplish this feat by allowing the operator to experiment with alternative architectural systems, each of which balances ecological variables in a different way. The storyteller, however, can only narrate possibilities that are contained within the software. As in all computer programs, there are only happy endings—endings that bear the inscription of the sustainable material and cultural processes initially authored by the programmer, who is, in this case, Fisk. In this sense, Icono-metric™ representation is both a learning tool and a tool that reproduces the intentions of the ecologists. It remains to be seen, however, if reproducing one's intentions in digital form can make any claims (without a biological existence) to being *sustainable*. This form of reproduction can, however, certainly make claims to being *critical* of the conventional technology promoted by the land grant network.

SUBLIME
REPRODUCTIONS

By gathering his intentions into the controllable, but non-biological, world of electronic space, Pliny Fisk was able to reproduce Blueprint Farm more brilliantly than he was ever able to do in the real world. Icono-metric™ and Biom-metric™ representations are not, however, material artifacts that produce space in any conventional sense. Although the material technologies that the ecologists put in place at the farm have effectively dematerialized, the literary and social technologies documented by Shapin and Schaeffer, and employed by Fisk, have successfully reproduced his intentions. Those intentions have become so visible, it is as if the material technologies abandoned or appropriated by others in Laredo were now operating very successfully indeed.

Not only was Blueprint Farm publicized in the professional architectural press, but it has also been publicized repeatedly in the environmental press and the popular press—including the *New York Times*.[17] Among advocates for sustainable architecture, the project has been elevated to the status of a cult object through its incorporation into overlapping literary and social networks. It is also arguable that the literary and social methods of fact production employed by Fisk have benefited from the geographic isolation of Laredo. The practical invisibility of the place has permitted "virtual witnessing" to take the place of experiencing the experimental apparatus itself. As a result, the literary and social reality of Blueprint Farm has gone unchallenged by material conditions.

Following the closure of Blueprint Farm, Fisk has become an increasingly popular speaker at national and international conferences that feature agendas concerning the concept of sustainability. Such visibility has helped to catapult CMPBS into national prominence as an ecological consultant to other architects who have ecological intentions (or clients with such intentions) but not the necessary research skills. For example, a few years after the closing of Blueprint Farm, the Center for Maximum Potential Building Systems was awarded a two-year research contract with the federal Environmental Protection Agency. CMPBS's work for EPA has been to expand upon the assumptions behind the Icono-metric™ life-cycle model and develop a tool with which to assess global ecological and technological resources.

My point in documenting the apparent and deserved successes of CMPBS is that the public space of architectural fact production is not the materials laboratory, or even the site of building construction—it is the rostrum. I am not arguing that this is necessarily a bad thing, it is simply the way things get done in this society. The witnessing of virtual reality in the glossy pages of journals, at conferences, in the classroom, or on the Internet is a "public show" that avoids the inconvenience of material reality. Leon Battista Alberti, in his *Ten Books of Architecture,* describes architecture as a "public show" of technological achievement.[18] One of the many ironies of Blueprint Farm is that its producers intended a "demonstration," or a "public show," in the manner recommended by Alberti from the sixteenth century, yet stuck the show in an isolated corner of an isolated town. For Alberti, the use-value of architecture has always been, to some degree, political rather than strictly material. Although architecture transforms the material realm, it also transforms the meaning of the public realm. The more sublime the technological achievement, the more satisfying is the "public show." The evidence in this case suggests that the invisibility of the selected site occurred more by circumstantial ineptitude administered by TDA than by conscious design. The evidence also suggests that the failure of Laredo Junior College to import witnesses to the site contributed to the local denial of the project's fact-ness. One possible explanation of this situation is that, at least for Hightower, the only group that had to be impressed and pacified by the spectacle of the farm was Texas Jews. These witnesses were both imported (in small but influential numbers) and virtually incorporated into the landscape of sustainability. If Hightower's fear of owing a political debt to the Jewish community over his Jackson endorsement was really the top priority, it didn't matter very much if other witnesses participated or not. In this view, what mattered was the reproduction of ties to the Jewish community, not the reproduction of knowledge, practices, or artifacts.

In the terminology of Henri Lefebvre, Blueprint Farm remained a "representation of space" rather than a "representational space." This distinction suggests that the farm never outgrew its conception as an a priori mental model of engineering efficiency. In lieu of becoming a "representational space," where geography is inhabited through association with the images and symbols of everyday life, life on this farm was an abstraction. Although such distinctions are hardly alien to modern life, the logical extension of Lefebvre's categories is to understand this project as a "public show," or theme park, rather than an inhabited farm.

A less harsh explanation for the failure of the ecologists to reproduce their sustainable machines is simply that they lacked sublimity in the eyes of local experts and farmers alike. David Nye best relates the concept of sublimity, first articulated by Immanuel Kant, to contemporary conditions. Simply put, Kant's definition of the *sublime* recognized that for eighteenth-century minds, the awesomeness of nature—its incomprehensible vastness—humbled men into a state of moral righteousness. Nye has recognized, however, that the phenomenon of sublimity has undergone a historical redescription in response to our technologically mediated world. In lieu of the *natural sublime,* which Kant documented, Nye has documented the recent appearance of what he calls the *technological sublime.* The technological sublime is not a communion of man with nature, as in Kant's world, but a communion of man with man—a celebration of the rationality of our species. In Nye's view, the technological sublime (experi-

enced at the Hoover Dam) relates to the future. In contrast, Kant's natural sublime (experienced at the Grand Canyon) relates to eternity.

Nye argues, in apparent agreement with Alberti, that the public experience of technology, at least in those special places like the Brooklyn Bridge, is
dependent upon the politics of perception. The social construction of powerful technological experiences is an intensely political activity that transforms
one's perception of reality. He warns us, however, that

> Like the Erie Canal or the Union-Pacific Railroad, the technological sublime gen-
> erates enthusiasm which the politicians seek to tap. The politically active citizen,
> so valued by republicanism, has been replaced [in American history] by the mes-
> merized spectator.[19]

> Unlike the natural sublime, which Kant defined as a source of value and as part
> of a process that led to a heightened awareness of transcendental reason, the
> corporate synthesis of various forms of the technological sublime sought not to
> enlighten, but to impress and pacify. The impulse to wonder had been subverted
> in magic and spectacle.[20]

In Nye's redescription of the concept, the technological sublime, unlike the
private contemplative experience understood by Kant, is generally a group
experience—a ribbon-cutting ceremony, a county fair, or a visit by foreign
dignitaries, for example. Nye defines the American experience of the technological sublime as the phenomenology of democracy, the collective experience of the heroic. He holds that the American imagination has wed (at least
until recently) democracy to technology. It is a phenomenon closer to political revivalism than to private transcendental experience.

If, as Nye suggests, the American imagination is addicted to the technological sublime, the sustainable processes imagined by the ecologists must have
bored young Laredoans, in particular, to tears. Although a sustainable technology might indeed be sublime—a mountain ridge capped by a line of wind gen-

erators, for example—it is more likely that sustainable technologies will be invisible—straw bales concealed by fly-ash stucco, or underground irrigation pipes, as was the case at Blueprint Farm. Of course, nothing could be less sublime than a composting toilet. Just as Texas architects interviewed in Chapter 6 demanded that "the good" be visually apparent in technology, nonarchitects demand that visual awe be apparent in technology. In contrast to the awesome army of gleaming tractor-trailers that cross the Rio Grande each day, the sustainable machines envisioned by Fisk are no competition. In the eyes of technological determinists, the slow plodding of natural cycles is, at best, a bore.

Boredom is, of course, easily exploited. Mystics and metaphysicians, priests and scientists, argue that knowledge emerges from "sublime, mysterious things."[21] In contrast, the geographer Henri Lefebvre argues that true knowledge emerges from work in the particular conditions of daily life. Lefebvre's *Critique of Everyday Life,* written in 1947, is an early attempt to radically secularize the redemptive claims of religion, art, and science. In Lefebvre's view, the external efforts of politicians (like Hightower), architects (like Fisk), priests, scientists, or other superhumans to impose new conditions of life in places like la Frontera Chica are necessarily a mystification of reality.

Although the sustainable and the technological sublime may not be mutually exclusive principles, they are inscribed in the same artifact only with great difficulty. Where sustainable technology is most interested in the repeated practices of everyday life, sublime technology is present only in the exceptional. It is this tension between the everyday and the exceptional, between craft and art, and between mechanics and science that is so easily exploited by the politics of perception. Politicians, and what critical theorists refer to as the "culture industry," are only too skilled at dispensing bread and circuses for the purposes of conquering markets and nonbelievers. In spite of Hightower's professed intentions to change the lives of small farmers, he was at least equally committed to exploiting Blueprint Farm as a technological spectacle for political purposes. Once local Jews were impressed and pacified by the Disney-like experience of witnessing real Israelis at work, the instrumental value of the farm was exhausted. The slow and inexact plodding of compost production or solar drying seemed to local citizens as somehow less than human and irrelevant to the trajectory of their lives. The technological sublime does not easily inhabit the commonplace.

These observations about the politics of perception require that I conclude this chapter by returning to the hypothesized relationship between the social construction of technologies and the social construction of places. First, this chapter has helped us to understand why the ecologists' technologies failed to be reproduced. It is not that the *sets of objects* themselves were rejected, but that the other two dimensions of technology proposed by MacKenzie and Wajcman—*knowledge* and *practices*—were never reproduced by locals themselves. In the definition of technology developed in Chapter 5:

> technology refers to what people *know* as well as what they *do*. Technology is knowledge. . . . Technological "things" are meaningless without the "know-how" to use them, repair them, design them and make them. That know-how often cannot be captured in words. It is visual, even tactile, rather than simply verbal or mathematical. But it can also be systemized and taught, as in the various disciplines of engineering. This indeed is the older meaning of "technology," one predating the use of the term to mean "hardware"—"technology" as systematic knowledge of the practical arts.[22]

The knowledge produced in the construction of the farm was not effectively taught by producers nor learned by receivers. The farming practices developed by the Israelis, or by the local farm managers, were never accepted by workers as *mestiza* and incorporated into their own operations at other locales. Local experts imagined that the objects would somehow reproduce themselves by immaculate conception. Without personal investment by local small farmers in the production of knowledge, and without their personal commitment to adopt radically new practices, the sets of technological objects left by the ecologists could not be locally reproduced.

This chapter has also helped us to understand Latour's proposal that "technological networks, as the name indicates, are nets thrown over spaces."[23] The metaphor suggests that there are multiple nets that tie places to other places, but that the links that do the tying are of various strengths. In this case, local farmers recognized that the links of both the Israeli network and the ecologist network were stretched farther and thinner than those of the land grant network—to which their lives had been tied for so long.

The sociologist David Brain has suggested that, just as the intentions of humans are *inscribed* in technological devices, such devices *transcribe* the places where they operate.[24] By substituting nonhuman actors, architects and politicians transcribe the scene of human events as it might have been played with all human actors. Although Hightower and Fisk had devised ingenious ways of transcribing the site and making their presence felt, in the end their links to the place were stretched too thin and broke, leaving small farmers alone and the objects to decay.

After the closure of Blueprint Farm and the election of 1990, Hightower cut his ties to agriculture. Both Fisk and the land grant network are, however, actively engaged in the redescription of the farm.[25] The project of the land grant network has been to redescribe the farm as an emblem of economic ruin and ensure that its reproduction will not occur. In the political climate of economic determinism that accompanies the new millennium, the most powerful redescription of "sustainability" imaginable is that which has been wreaked upon Blueprint Farm—that of an *economic* failure. In contrast, Fisk's project has been to *transcribe* the place electronically and reproduce the landscape of sustainability from the rostrum. The possibilities for multiplication in this nonbiological medium are, of course, vast. It will be some time, however, until anyone will be able to gauge how the brilliant, and sublime, images produced via communication technologies contribute to the construction of sustainable or even durable places. For the moment, however, electronic space seems an ideal locale in which to preserve and reproduce utopian intentions.

EIGHT PROPOSITIONS

In the preceding chapters I have presented an interpretive dialogue that has freely mixed three types of interpretation: the reflections of those on the scene, the theories of distant authors, and my own analysis. I recognize, of course, that the dialogue as a whole has been entirely my own construction because it is I who have selected locals to speak, it is I who have selected the theories of others to cite, and it is I who have woven this text. In the strictest sense, then, this study has been a monologue. I believe, however, that the empirical and discursive nature of my text has served to illuminate the complex social construction of architectural *intentions, interventions, receptions,* and *reproductions* in a way that a purely analytical text could not. As a result, some readers may accuse me of taking Clifford Geertz's admonition to write "thick descriptions" a bit too seriously.[1] The thickness or complexity of this description demands that I recognize the value of those other interpretations that will not agree with my own. Allowing for competing versions of reality, however, does not relieve me of the responsibility to summarize and generalize my own. That is the purpose of this concluding chapter. I will do so in three parts: First, I will offer eight propositions that summarize events that took place on the farm. These propositions are stated locally—meaning that they refer only to the specific conditions of Blueprint Farm. The second section of this chapter will rely upon these eight propositions to construct a *nonmodern* position that will avoid the antinomy of technology and place that is lodged in the worldview shared by moderns and postmoderns. This nonmodern conceptual structure, just like its modern counterpart, offers positive and negative valuations of the concepts technology and place. I will refer to these positions as *regenerative architecture* and *radical nihilism*. The third and concluding section, "Eight Points for Regenerative Architecture: A Nonmodern Manifesto," will renovate Kenneth Frampton's critical regionalism hypothesis by generalizing the eight local propositions drawn from the experience of Blueprint Farm.

SUMMARY

PROPOSITIONS

1. Blueprint Farm is a locale, or social setting, for the continuing struggle to achieve human liberation and self-realization through technology.

The events that add up to becoming Blueprint Farm can only be understood historically. In my reconstruction of the history of local space, I argued that the dislocated farmworkers, in whose name the demonstration farm was developed, seek increasingly abstract vehicles of liberation from the place-bound conditions of their servitude. They are not alone in this project, but have joined an emerging form of transnational subjectivity based upon transience as a way of life. In this sense, local farmworkers are thoroughly modern because they receive technology as a liberating force and receive transience as a liberated condition. This is not an argument that the modern condition of homelessness is a good or creative one. It is only a presentation of how displaced farmworkers tend to see their own situation.

Local class relations and patterns of property domination in la Frontera Chica have operated more like Central American models of aristocratic agrarianism than like the communal New Mexican model of *ejido,* or the Jeffersonian model that has been operational in most of the United States. In the communal model of New Mexico, even privately owned lands were understood as public resources. In the Jeffersonian model, the land is associated with independence, self-realization, and social democracy. In contrast, the contemporary Central American agrarian model associates the land with exploitation and hierarchical social structures. In the view of Laredoan farmworkers, then, it is not surprising that the implements of "sustainable" agriculture were only symbols of servitude to the land and its owners.

In the current situation, farmworkers see trucks, rather than agricultural implements, as the more potent vehicle of liberation and self-realization. A truck would, of course, be a useful tool to a creative farmer who wished to create a *mestiza* distribution system to serve those conscious citizens who don't shop at H-E-B and who live between the nodes of the dominant technological networks. It is more likely, however, that displaced farmworkers who are fed up with the drudgery of stoop work would choose to use their new trucks as implements of commerce rather than agriculture. In other words, a truck—preferably a big truck—would enable the underemployed to haul rather than hoe. While it is impossible to generalize, the evidence suggests that, given the choice, most displaced farmworkers would rather *participate* in the economy of exchange. As metaphors of mobility and as useful tools, the trucks desired by local workers tend to abstract their lives and loosen ties to those traditional places that they have already left behind.

Implicit in the dependence upon trucks to achieve liberation and self-realization is the concept of technological determinism—the assumption that there is an irrefutable logic to the ubiquity of those technological objects that enable a way of life. The logic contained in trucks, for example, seems to prohibit places like Laredo from being lived other than transiently—even for those who have spent their entire lives there. In this sense, la Frontera Chica has been, at least for marginal farmworkers, a social setting through which to move on the way to a better life.

2.   There is no available constellation of ideas in Laredo upon which the concept of sustainability might hang.

The intentions of the five technological networks identified as operating at Blueprint Farm were thrown into historical conditions. Jews, local experts, the Hightower network, the ecologists, and the land grant network each had distinct visions in mind for Texas agriculture in general, and for la Frontera Chica in particular. What the ecologists had in mind, however, proved to be the greatest stretch of all. The culture of la Frontera Chica lacks an interpretive paradigm, other than the concept of *mestiza,* that might have proved helpful to local farmers in understanding the radical intentions of the ecologists, or to the ecologists in relating their intentions to local traditions and modes of understanding. Although *mestiza* culture has very little to do with ecologism per se, it has very much to do with the production of alternative space. Tragically, those who developed the farm were either unaware or not supportive of the noncapitalist agrarian practices of Hispanics in New Mexico. The concepts of *ejido* and cooperation might have served as a point of attachment for the concept of sustainability to indigenous paradigms of interpretation.

Much to the surprise of the ecologists, the market-driven intentions of the Israelis and the land grant network were far more familiar to local farmers than the organic constructs of social ecology. Although Jim Hightower is a master of populist rhetoric, he failed to connect his own political interests to the everyday concerns of displaced farmworkers. Pliny Fisk suffered a similar failure in translating the interests of social ecology. Under normal conditions of conversation, he can be hard to follow, but, put into the context of everyday life in la Frontera Chica, he becomes unintelligible. In this case the producers of the farm found no local ideas or practices to which the intentions of the ecologists could be hung. The reverse interpretation is also warranted— that the ecologists, like their collaborators, acted ex nihilo, meaning that they entered la Frontera Chica with a priori definitions of sustainability that were imposed upon local practices. Those who were intended to benefit from the production of space were excluded from its very conception.

3. The producers of Blueprint Farm mistook their constructed interests for their intentions.

Rather than understand their activity as the desire to live differently, the participants in four of the five networks that produced Blueprint Farm imagined that the objects themselves would alter the conditions that concerned them. Another way to say this is to hold that local experts, the land grant network, the Israelis, and the Hightower regime only expected the lives of *others*— meaning displaced farmworkers—to change. In this sense, these producers did not *inhabit* their work--they were too busy promoting their own ideological interests. All but the ecologists, then, operated as blatant technological determinists because they assumed technological objects to be autonomous—that is, independent of human knowledge and human practices. In other words, they mistook *actual objects* for *intentional objects.*

It was a surprise to all but the ecologists that the objects would not operate without an accompanying transformation of local knowledge and practices. The *comportment* of the ecologists toward their work was different. Where their collaborators always intended to build objects, the ecologists intended to build machines—processes that would link human institutions to evolving ecological conditions. In the end, however, even the ecologists confused their ideological interests with their intentions, and they—like any other business with a payroll—moved on to other projects. If the others were *hard* technological determinists, the ecologists were *soft* determinists.

4. The regenerative technologies of Blueprint Farm resisted the gravitational pull of the centers of calculation.

I have argued that the regenerative technologies proposed by the ecologists, such as wind-towers, are inherently more democratic than are universal appliances like compression air-conditioning. Such transparent technologies make visible the local labor conditions under which technological objects are produced. Regenerative technologies also measure the performance of systems in both qualitative and quantitative terms. For example, a river breeze may be ex-

actly ten degrees cooler than the ambient air, but the experience of that coolness defies quantification as "degrees." It is precisely this anarchic quality that
renders the technologies constructed at Blueprint Farm resistant to incorporation by large technological networks. Because regenerative technologies cannot be predicted and controlled with universally applicable techniques, they
resist what Latour describes as the "centers of calculation." These centralized
locales of data and profit accumulation operate most efficiently when universal abstractions can be manipulated without the interference of confounding
local variables. In contrast to the abstract criteria that enable appliances, regenerative technologies magnify, rather than erase, local labor and ecological ir-
regularities, thus becoming resistant to centralized production and marketing
strategies. Under such conditions no "black box" can fully close.

5. *The regenerative technologies of Blueprint Farm lacked sublimity.*

Those who produced Blueprint Farm, all of whom were technological determinists to one degree or another, perceived a lack of sublimity in the low-
tech structures installed by the ecologists. Although the nontraditional appearance of the shade structures, wind turbines, and packing sheds was of
some political value to Hightower and local experts as a spectacle, in the end
that spectacle was not grand enough to pacify all of the farm's detractors.
Much to Hightower's consternation, the ecologists seemed to subvert the
spectacular possibilities of the farm. Rather than finishing the big-ticket gadgets like solar/zeolite refrigerators, the ecologists devoted their energies to
such earthy projects as composting and community gardening. Of course, it
is these frustrated outreach efforts that were potentially most important to
local farmers. But for Hightower and Laredo Junior College administrators,
these low-tech toys only created problems of perception that had to be made
to disappear. Where the majority of producers found political value in the
sublime dimension of technology, the ecologists found value in the regenerative dimension of technology. Jews and locals could not be made to for-
get their troubles by pondering the plodding cycles of compost production

or zeolite refrigeration. They could be pacified, however, by witnessing real Israelis produce agricultural exotica. The conditions of technological sublimity require the absolute closure of black boxes—a condition that is inconsistent with the demands of participatory democracy.

6. *The closure of the regenerative technologies employed at Blueprint Farm came, not by incorporation into spaces, but by suppression.*

"Closure" is a moment in the social construction of technological systems that is often mistaken for the moment of invention. It is a moment of agreement when the majority of producers decide how knowledge, practices, and objects will be integrated into a mode of production. It is the moment when the lid of the black box is closed. Given the fundamentally opposed intentions that existed between the ecologists and the four other technological networks at Blueprint Farm, it is hardly surprising that closure came, not by the inclusion of these systems in spaces controlled by the others, but by their suppression. Jews, local experts, the Hightower regime, and the land grant network all found that more problems were made to disappear by suppressing the technologies offered by the ecologists than by supporting their continued development. Jews found that the embarrassing struggle over authorship of the farm disappeared with its closure. Laredo Junior College administrators found that the unexpected drain of scarce resources disappeared with the closure of the farm. The land grant network found that the challenge to entrenched bureaucratic categories disappeared with closure of the farm. Only the ecologists and forgotten local farmers would have benefited by continued public access to the contents of the black box.

7. *The technology and place of Blueprint Farm are different things, but the social processes of their construction are dialogically related.*

This proposition is the hypothesis that was introduced in Chapter 3. There I concluded that the dimensions of place—*location, locale,* and *sense of place*—are largely congruent with the dimensions of technology—*knowledge, practices*, and *sets of objects*. From the reconstructed historical conditions of la Frontera Chica, I hypothesized that the relationship between technologies and places is *dialogic*. I intend by this term the conversational, or hermeneutic, quality of human agreements that *take place*, as opposed to those agreements we understand as purely mental constructs. As I defined my terms in Chapter 3, such a *dialogic* relationship is distinct from the *dialectic* opposition of principles understood by traditional Marxism. Where a dialogic relationship will either converge slowly toward a common horizon of meaning, or diverge, a dialectic relationship will either transcend into synthesis (in its Hegelian form) or remain in a symmetrical state of creative opposition (in its critical theory form). A dialogic relationship is not negative or transcendental, nor can it be reduced to common conversation.

The four chapters that followed Chapter 3, "A Local History of Space," have supported that dialogic hypothesis. They have demonstrated that, in this case, the divergence of the Israelis' high-tech machines and the low-tech machines of the ecologists was situational, or linked to the particular material discourse of Blueprint Farm. The knowledge and practices of competing technological networks were carried in the space-occupying objects constructed at the farm. For example, the appliances imported by the Israelis (computers, pipes, and valves) produced a very different place than did the regenerative technologies (wind-towers, compost cookers, and straw-bale walls) constructed by the ecologists across the DMZ. I must stress that, in spite of the fundamental opposition that developed between these two ways of knowing and doing, *the divergent outcome might have been different*. For example, although the technology of Fertigation™, as used by the Israelis, lends itself conveniently to chemical farming and centralized planning, after the Israelis departed, the ecologists demonstrated that this technology lent itself to organic farming equally well. Fertigation™ is not, in itself, an inherently undemocratic technology. This example suggests that, without the burden of Jim Hightower's political debt to Jews, without the burden of a few particularly prickly personalities, or with different management at the local H-E-B grocery store, it is quite possible that the dialogues that took place at Blueprint Farm might have been *convergent* rather than *divergent*. The history of space, and not the technological objects themselves, drove this particular dialogue to an ironic conclusion. To generalize this logic would be to suggest that altered political conditions might yet lead to a constructive, hermeneutic dialogue and a successfully constructed locale.

Technologies enable and disable places. By suggesting that technologies *enable* places, I am not proposing that technologies *determine* places. Nor am I arguing the converse, that places determine technologies. Technological determinism and environmental determinism are equally problematic discourses, which I reject. My point is that places and technologies are enabled, and disabled, by the same dialogic process of reaching social agreements about how to live.

In the battle for imaginative supremacy to describe the politically useful concept of *sustainability,* the ecologists and the marketeers in this case failed to reach common agreements concerning technology that might have enabled a place to emerge. An analogy would be to argue that New York City, for example, was enabled by the human agreements inscribed in the steel frame and the elevator, or that Houston was enabled by the agreements inscribed in the automobile and compression air-conditioning. Likewise, I am arguing that Blueprint Farm was *dis*abled by the failure of those involved to agree about how food can be sustainably produced in a semiarid ecosystem. The point here is that it was not the technologies of Blueprint Farm that failed, nor were there inalterably opposed principles inscribed in them. Rather, it was the human agreements to incorporate those technologies in the practices of everyday life that failed. In the end, the ecologists were unable to mobilize support for their version of sustainability within the four competing technological networks. As a result, a distinctly sustainable place never showed up.

8. *Blueprint Farm can be understood as a critical place, but not as a sustainable or regenerative place.*

This final proposition requires a more lengthy treatment so as to set the stage for the conclusion to this study. According to Henri Lefebvre, social spaces are made distinct by the particular qualities of the mode of production that is employed to construct them. Just as market-determined spaces (the shopping mall, for example) have certain characteristics, or theocratic spaces (such as Mayan temples) have certain other characteristics, sustainable or regenerative spaces would have equally distinct characteristics. They would be, among other things, islands of diminished entropy. Blueprint Farm was an attempt to investigate what those "other things" might be. In this view, the places produced by contemporary market forces are no less distinct than those produced by Mayan temple-builders. In each case the objects and spaces of transformed nature bear the inscription of the respective mode of production. What distinguishes Blueprint Farm from the market-driven spaces that surround it is not, then, so much an aesthetic issue as an issue about *how* nature should be transformed. It is an issue of political economy.

To develop this notion of the politics of space, it will again be helpful to return to the concepts developed by Bruno Latour. He has argued that the technological networks that dominate market economies (like the land grant network identified by Jim Hightower) operate as "nets thrown over spaces."[2] His metaphor suggests that between the links and the nodes of the market, there are no trucks, no roads, and no used car lots, not even any television sets. It is not that the spaces circumscribed by technological networks are empty, or uninhabited, it is just that they are *unconnected* to the web of human agreements that are woven into technological networks. Technological networks are essentially spatial, but no matter how tight the weave of the net, some spaces are left unconnected to the centers of calculation. They are differentiated spaces. David Harvey, for one, takes exception to the proposition that marginal spaces might somehow avoid "the dialectic."[3] My point here, however, is not to promote the creation of utopian spaces that consciously avoid participation in the public discourse, or avoid telling us how to get from here to there, but to argue for the development of spaces where alternative discourses might begin, or might be preserved. It is these unconnected, or differentiated, spaces that are available to other, or noncapitalist, modes of production. I will refer to those spaces that transform nature in some non-market-driven mode as *critical places.*

Historically, a list of such critical places might include Walden Pond (Henry David Thoreau's agrarian retreat near Boston), Arcosanti (Paolo Soleri's "arcology" near Phoenix), Wounded Knee (the martyred locale of the American Indian Movement), and Llano del Rio (the socialist commune that Mike Davis portrayed in *City of Quartz* as an alternative history for the city of Los Angeles).[4] Of course, the Hispanic concept of *mestiza*—the cultural interstices of the borderland—is a very positive example of a critical place. My point is that *critical places are inscribed with the suppressed history of minority agreements.* They are socially described spaces that defy the zeitgeist, or dominant spirit, of the time. In our time, that spirit is relentlessly driven by the forces of economic determinism—the notion that one's very existence must be justified in the reductive terms of the economy. Thus any place that emerges through alternative modes of production, or alternative discourse, is, in my view, a critical place.

Alvaro Lacayo, the Laredoan labor activist, grinned widely when he said that "the farm stands as a memorial to what can be, not here, but maybe in the Gobi Desert, or in Chihuahua."[5] In his utopian anticipation of what might be, Lacayo has nominated Blueprint Farm to occupy a spot on the illustrious list of critical places. Lacayo is not alone among the producers of Blueprint Farm in memorializing the intentions inscribed there. Celia Juárez sighed that "we just reached too far,"[6] and Dr. Jiménez said with the deflated hindsight of a pragmatist, "it was a dream."[7] These participants recognized the farm to be a precinct unconnected to the ordinary world.

In Lacayo's nomination, he used the troubling word "memorial" to describe an essential characteristic of the critical place. I find the term troubling because humans do not inhabit memorials. They are designed to serve the remembrance of lost, or as yet unmaterialized, persons, events, or possibilities. Memorials serve absent presences. As machines of remembrance, memorials can preserve possibilities, but they are not in themselves life-enhancing. Memorials exist outside, or after, life. They are *extra*ordinary and distinct from what Lefebvre has described as a "monumental space"—a space that "offered to each member of a society an image of that membership, an image of his or her social visage."[8] Where Lacayo's privileged space memorializes causes lost by the few, Lefebvre's privileged space monumentalizes a collective mirror of faith. These are very different references.

With regard to Blueprint Farm, Lacayo's desire to preserve possibilities and the project of life enhancement require a bit more investigation to clarify the nature of critical places. In Chapter 4, I suggested that the ecologists were seduced by the purity of failure. In the logic being developed in this conclusion, however, the production of memorials critical of the dominant mode of production can hardly be considered a failure per se. Many consider such memorializing to be a heroic, or at least instructive, act. I will argue that, for Fisk and his collaborators, Blueprint Farm has been a highly successful critical operation. By *critical,* I mean that the suppressed intentions of the ecologists were not so much to inhabit a place in the Heideggerian sense of *dwelling* there, as to deposit a diagram that would critique how others farmed in *all* semiarid locales and at *any* time. The instrument of that critique was technology. My point is that successful critiques need not be sustained, or reproduced. The ecologists in this case, in spite of their regenerative intentions, inadvertently fell back into modes of operation that are more consistent with anarchist utopian traditions, or the historical fantasies of Marxism, than with postmodern ecologism. By utopian, I mean that their intentions were not really place-specific, nor did they intend the project to solve immediate social and ecological problems. Rather, the property of Laredo Junior College was used instrumentally to create a critical display of technology that described a future possibility.

The utopian possibilities of Blueprint Farm have been marginally preserved, or memorialized, by its conversion to a "museum" dedicated to the interpretation of ecological conditions. The reproduction of a few isolated examples of Fertigation™, the growing popularity of rainwater harvesting in the Southwest, and the surge in straw-bale construction all over America are to some degree reproductions of the ecologists' intentions. As powerful as these appropriated and reproduced artifacts may be, however, Fisk's Iconometric™ mode of representation has done the most to preserve the utopian possibilities of the farm. These computer technologies have not, however, been reproduced in any social or material sense. Rather, they have been preserved in hyperspace as utopian intentions, divorced from the material conditions of everyday life and everyday places. Although the manipulations of electrons in hyperspace might be a critical activity, and utopian in its intentions, it cannot in itself be a life-enhancing practice. Computers operate in a nonbiological world. The point that I want to stress here is that there is, in both the material and electronic worlds, a paradigmatic tension between critical practices and regenerative practices. I mean by this that to critique the conditions of everyday life from a distance is a very different activity than to regenerate the conditions of life as a participant. In this sense, Blueprint Farm should be, in the end, understood as a critical place and not as a regenerative place. To be clear, I am proposing that we value highly a critical blueprint for the future, but not as highly as the resolution of problems experienced by those transients who now eke out a living in Laredo.

Critical theorists, Theodor Adorno, for example, would argue that it is precisely the distance of Lacayo's memorial from the conditions of everyday life that renders it critical of market-driven practices and therefore helpful in the project of exploring alternative futures. In opposition, Lefebvre would argue that it is precisely the distance of the memorial proposed by Lacayo from the material conditions of everyday life that renders it helpless in the political project of life enhancement. What separates these two positions is the insistence of Adorno and critical theory on an alienated critique of conventional practices, and Lefebvre's insistence on participating in normative practices that are life-enhancing. Even David Harvey, whose allegiance to the principles of Marxism is never in doubt, recognizes that "The ecological critique of socialist 'productivism' is here helpful, since it forces Marxists to re-examine the problematics of alienation."[9] In this passage, Harvey (like Lefebvre) acknowledges that Marxists have clung far too narrowly to the implications of alienated labor and have thus ignored the implications of our alienation from nature. Such myopia has led to a negativism that is, in our current situation, less than helpful. Where traditional Marxism is relentlessly alienated and negative, Lefebvre manages to construct a socialist position that is positive, life-enhancing, and, I believe, a precursor of regenerative thought. In lieu of the alienated utopianism of Lacayo and Fisk that reveals itself in the case of Blueprint Farm, and which avoids the messy realism of everyday practices, regenerative architecture would inhabit the everyday interstices of local/global networks and participatory democracy. In other words, regenerative architecture in la Frontera Chica would be a *mestiza* practice.

At the most fundamental level of epistemology, Marxism rests upon the dialectic opposition of knowing subjects and known objects. It is, of course, just this modern bifurcation of the world as alternately human and nonhuman that regenerative thought seeks to replace. The descendants of Martin Heidegger argue that, without moving beyond the Cartesian notion that human subjects operate in mental space that is distinct from the material space inhabited by trees and trucks, it will be impossible to participate in the balancing of the sociobiological processes upon which life, liberation, and self-realization depend.

As Latour has argued, and as the evidence in this case has demonstrated, the radical alienation of subjects and objects, or of human interests and technological networks, is a myth of modernity that is inconsistent with the reality of our social practices. Whether we acknowledge it or not, the preontological practices inscribed in places are always already directed toward the material world. To preserve possibilities by memorializing objects is historically helpful, but will not, in the end, regenerate an increasingly imperiled world. What is needed is life-enhancing practices, or regenerative technologies, that unite

quasi-subjects and quasi-objects in a single world. An uninhabited "demonstration" might be critical, and historically instructive, but it cannot sustain or regenerate life in concrete ways. I will finally hold that the concepts of sustainability
and regenerative architecture might be *lived,* but cannot be *demonstrated.* A
demonstration of technology can only be critical, or utopian. This, then, was a
principal misconception that guided the development of Blueprint Farm from
the beginning. One cannot demonstrate how others should live their lives. To
do so remains, not unlike the well-intended schemes of urban renewal in the
1960's, within the modernist legacy of grand plans.

In the introductory chapter of this study I suggested that **THE NONMODERN**
Kenneth Frampton's critical regionalism hypothesis would **THESIS**
provide a way out of the dialectic opposition between
place and technology that has been constructed by modernist thought. My project has been to understand how the reconstructed
conditions of Blueprint Farm might contribute to the extension of Frampton's
hypothesis. On the basis of the foregoing summary, my conclusion is that critical regionalism must be removed from its roots in dialectic logic and critical
theory and grafted to a *dialogic hermeneutic* construct. In other words, I am
proposing to transplant Frampton's hypothesis from an alienated logic dependent upon transcendental or oppositional interpretations of reality to a
conversational logic or relations dependent upon emergent and collective
interpretations of reality.

The rationale for the repositioning of critical regionalism is one of methodological fit. As Fredric Jameson has observed, critical regionalism is neither
modern nor postmodern in its basic assumptions.[10] The positive position of
critical regionalism toward the concepts of both technology and place is indeed misplaced among modern assumptions, in which human subjects and
nonhuman objects occupy alienated worlds. Critical regionalism is equally

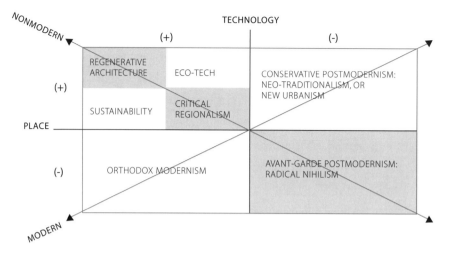

Figure 8.1. Alternative theoretical positions with regard to the concepts place and technology.

misplaced among postmodern assumptions, in which the values related to technology and place are, as I argued in Chapter 1, simply inverted. The positive synthesis of the modern dualism requires a set of axiomatic assumptions in which humans and nonhumans *inhabit* a common world that is simultaneously mental and material. What will be helpful, then, is to clarify the position of Frampton's hypothesis as a nonmodern alternative to our current situation. The term *critical regionalism* is, however, no longer appropriate to communicate nonmodern assumptions because the term *critical* relates the reader back to the modern assumptions of critical theory. For this reason I am suggesting that we renovate Frampton's terminology and rename the emerging hypothesis as a proposal for *regenerative architecture.* Figure 8.1 will be helpful in mapping alternative theoretical positions with regard to the concepts of technology and place.

This figure suggests that the axis constructed by nonmodern assumptions can resolve the modern/postmodern opposition in either a positive or a negative synthesis. Before I offer a fuller account of regenerative architecture it will be helpful to review briefly the positions identified in each of the four quadrants of Figure 8.1.

In Chapter 1, I argued that modernism is virtually synonymous **The Modern Axis**
with the positive use of technology. I also relied upon John
Agnew to demonstrate the modern antipathy toward the traditional concept of place. If the reader will recall, Agnew argued that modern
thought has devalued the concept of place in two ways. First, moderns have
conflated moral concepts and physical places; and second, moderns have
constructed a teleology that assumes the course of history to be directed
away from place-bound social hierarchies and toward the liberative conditions of transience and Cartesian space. I also argued that, for modern architects, Le Corbusier's *purist* architecture of the 1920's is perhaps the best exemplar of the modernist position.[11] His Plan Voisin for Paris, illustrated in Figure
8.2, is an archetype of early modernist ideology.

In my view, it is pointless to argue, as do conservative postmoderns, that
the geography presented in this image is not a place.[12] I suspect conservatives mean that Le Corbusier's Paris is a geography found to be so offensive,
so threatening to traditional modes of being, that they refuse to recognize its
very existence for fear that it will contaminate those real (read "traditional")
places that remain to them. Rather than deny the existence of alternative
modes of being, however, it seems more productive to argue that Le Corbusier's Paris is a place made distinct by a particular mode of production
working upon nature through the medium of modern technology. Rem
Koolhaas has argued that Le Corbusier's use of technology was both instrumental and representational (characteristic, in my view, of liberal capitalism):

> For Le Corbusier, *use* of technology as an instrument and extension of the imagi-
> nation equals *abuse*. True believer in the *myth* of technology from the distance
> of Europe, for him technology itself was fantastic. It has to remain virginal, can
> only be displayed in its purest form, a strictly totemic presence.[13]

In the terms that I used to analyze Blueprint Farm, Le Corbusier's Plan Voisin
was *enabled* by the apparatus of modern technology. It is not that Le

Figure 8.2. Le Corbusier's Plan Voisin for Paris. Le Corbusier, from *Urbanisme* (Paris, 1922).

Corbusier's Paris is a nonplace, as romantics would have it, it is simply a non-traditional place envisioned by a technological determinist. To be clear, I want to emphasize that there is no necessary correlation between the concept of place and an old-style community. Rather, all places reproduce the manner of their making.

If the modern thesis is exemplified by the early work of Le Corbusier, what of the postmodern antithesis? I have argued that conservative postmoderns have merely inverted the dialectic opposition between places and technologies that has been constructed by moderns. Put more directly, this is to claim that *the postmodern is only antimodern.* In lieu of creating places consistent with the doctrines of technological determinism, as did moderns, conservative postmoderns would create places consistent with the doctrines of environmental determinism. But rather than revisit this discussion in purely abstract terms, it will be more instructive to consider an exemplar of the postmodern inversion of the modern dualism.

According to Kurt Anderson, Seaside (shown in Figure 8.3), the much-publicized icon of neotraditionalism, or new urbanism, designed by Andres Duany and Elizabeth Platter-Zyberk (DPZ), is a new "planning paradigm" for America. The paradigm so lauded by Anderson relies upon two codes: The first, the urban code, defines property, density, circulation, and building-use types. Most planners agree that the simplicity of DPZ's urban code has provided a much-needed critique of the automobile culture inscribed in conventional suburban planning models. The second, the architectural code, is, however, a far less convincing recuperation of presuburban typologies. This code, rather than promoting environmentally responsive design strategies, promotes the reproduction of traditional architectural motifs for their scenographic value. Anderson, far from being apologetic for mere image-making, delights in the recuperation of what he describes—using the language of former President George Bush—as "a 'kinder-gentler' model for the future which portends 'a return to hearth and home.'"[14]

Figure 8.3. Seaside, Florida, by DPZ, Andres Duany and Elizabeth Platter-Zyberk. Photograph by Wes Henderson.

In response to Seaside's critics on the left, Anderson retorts:

Of course Seaside is fundamentally an exercise in nostalgia, seeking (like practically every other suburb in the country) to indulge middle class America's pastoral urges. The miracle is that (unlike practically any other suburb in the country)
it manages to conjure the good old days impeccably, solidly, jauntily, even
profoundly.[15]

The claim to profundity does not obscure the reactionary content of
Anderson's message. In this view, acts of "nostalgia" need only be performed
"impeccably" to legitimize their mission of reversal.[16] Explicit in his admira-
tion of Seaside is a "return" to the presumed grace of "the good old days" found
inscribed in the forms of "hearth and home." I will argue, however, that it is
not the forms of Seaside that are so admired by new urbanists. Rather, it is the
structure of social relations represented in the classical-vernacular forms ordered by DPZ's architectural code that is so admired. The simulation of traditional places is a thinly veiled attempt to recoup the grace and privilege of
those who would reside at the top of traditional social hierarchies. Anderson,
like the moderns who so annoy John Agnew, conflates social and architectural typology.

In the realm of reactionary cultural politics, however, Andres Duany
is not to be outdone. Recognizing in them the role of the long-lost feudal
"prince,"[17] Duany heroizes

developers: they are the best kind of client, because they have a great deal of
power over place, power equivalent to the aristocracies of the past. . . . History
shows that you have to concentrate power to achieve decisive physical form, and
developers are currently the people who hold such power in the United States.[18]

Well, so much for the Enlightenment project of liberation from the place-
bound bondage to the prince. In Duany's world of "decisive physical form,"

gone, too, are the democratic aspirations of progressive postmoderns. At Seaside, the simulation of traditional places is a setting for the reenactment (and cynical marketing) of traditional relationships. The difference, however, is that in this social setting everyone who can pay the price identifies with the prince of real estate, not with the banished peasant. Postmodern place, as an embodied set of conservative social relations, has edited space so as to exclude those unfamiliar Others. In the end, Seaside is only antimodern because it accepts the modernist reification of moral codes, yet intends to reverse the teleology of modernist history.

Figure 8.1 identifies the position that negatively values both **The Nonmodern Axis** place and technology as *radical nihilism*. To introduce a new term in the conclusion of this study may prove to be a source of distraction. However, the concept of radical nihilism will serve to clarify, if only by inversion, the theoretical focus of this study. Simply put, radical nihilism reverses the assumptions of critical regionalism and regenerative architecture.

The traditional concept of nihilism that derives from Friedrich Nietzsche argues that nothing exists that is knowable, or that can be communicated. For Gianni Vattimo, this strong version of nihilism ultimately devalues its own claim because, if *nothing* is knowable, the nihilist claim itself must be thrown on the groundless quicksand of relativism. Out of this modern impasse, Vattimo concludes that our only recourse is to radicalize, and thus disable, the modern will to Truth.

Vattimo's proposal for a radical nihilism holds that modernity (and history itself) have ended, not because of the apocalyptic trajectory of modern technology, but because the ideology of progress has rendered the culture of the new to be routine. If we wish to escape the clouded future of modernity and the horrors of its totalizing projects, our only escape route will be to radicalize the very nihilism inherent in modernity itself. The effect of this move will be to secularize the possibility for utopia as a weakened opportunity for authentic being.[19] Vattimo's project of secularization is the same process by which the Christian myth of redemption or the modern myth of utopia is demythified and emptied. It is to resign oneself to metaphysics, not by overcoming it, or by overcoming one's concern for its implications, but by somehow swallowing it and getting beyond the need to rationalize the world. It is an attitude analogous to recovering from an illness. If modernity is characterized as a period of overcoming, one cannot overcome "overcoming." One can only "dissolve modernity through a radicalization of its own innate tendencies."[20] Radical nihilism is, then, a thoroughly nonmodern position because it rejects the assumptions of moderns and conservative postmoderns alike. This brand of postmodernism is, however, quite distinct from the conservative position exemplified by the new urbanism of Duany and Platter-Zyberk.

Gevork Hartoonian has expanded upon Vattimo's thesis of secularization. Unlike Martin Heidegger and those who wish to return to a "unity of means and ends," as Hartoonian puts it, he does not lament the estrangement of organic "kulture" from the rationalized process of "thinking and doing." Hartoonian argues that "the mechanization of production has made it impossible to transfer tradition, including the craft of architecture, without subjecting it to the process of secularization that has supplanted the Christian idea of redemption with the idea of progress."[21] For Hartoonian, the process of secularization is the process of substituting a progressive tectonic for a redemptive one.

The most visible exemplar of this secularized logic is the work of the Dutch architect Rem Koolhaas. At a 1995 urban design conference in Singapore, Koolhaas shocked Asian architects (who expected the support of Western colleagues for their efforts to resist globalization through historic preservation measures) by insisting that "Instead of resisting this globalization, we should theorize about it. Perhaps we have to shed our identities. Perhaps identity is constricting us." In that forum Koolhaas argued, without a trace of irony, that once precious local identities are lost, as they inevitably will be, only "the generic" will remain. Koolhaas concluded, "...brand names are less important than the generic. The city is now a plain envelope. Any regret about the loss of history is just a reflex" that serves no useful purpose. Rather than mourn the loss of authentic places, argues Koolhaas, we should recognize that life in radically dislocated cities like Singapore is simply more interesting than the traditional places that are inevitably lost. In terms reminiscent of those advocated by Vattimo, Koolhaas has embraced the end of history and the radical secularization of sacred ground in the name of generic place. For radical nihilists, such as Koolhaas, the "culture of congestion" is preferred to the "urbanism of good intentions" promoted by Duany and Platter-Zyberk.[22]

If Koolhaas devalues the traditional concept of place, how does he value technology? In his 1978 book, *Delirious New York*, Koolhaas has famously argued in favor of "the technology of the fantastic," which he likens to Coney Island, and "a permanent conspiracy against the realities of the exterior world."[23] In his embrace of the irrational, Koolhaas rejects Le Corbusier's interpretation of the metropolis as sanitized and Cartesian in favor of Salvador Dalí's chaotic and antifunctionalist interpretation. Rather than employ pure technology as an end in itself—as he has accused Le Corbusier of doing—Koolhaas would employ technology as "the most rational possible instrument at the service of the most irrational possible pursuit."[24]

On the visual evidence of Koolhaas's work, such as his urban design proposal for Yokahama, Japan, illustrated in Figure 8.4, and on the evidence of his statements cited above, I will argue that his position exemplifies the negative nonmodern synthesis of the modern/postmodern dualism. Koolhaas rejects the redemptive qualities found in technology by moderns, and he rejects with equal vehemence the redemptive qualities of traditional places imagined by conservative postmoderns. Like Vattimo, however, he would not see his position as particularly negative, antiplace, or antitechnology. Rather, it is that technology and place are simply devalued concepts. His is a proposal for "weak being" which encourages us to accept the infinite interpretableness of reality. Once we do, argue the radical nihilists, the strong structures of modernity (like technology, utopia, and truth), as well as the sacred structures of conservative postmoderns (like place, territory, and authority), become weakened by the relativity of weak being. The radical nihilists argue that generic place

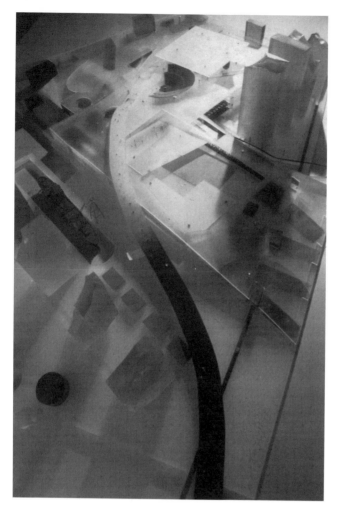

Figure 8.4. Model study of an urban design proposal for Yokahama, Japan. Courtesy of Rem Koolhaas and the Office for Metropolitan Architecture (OMA). Photograph by Hans Werlemann © 2000.

and instrumental technology, rather than being fallen, or dark, angels, don't seem so bad. In my own view, the radical nihilists offer a coherent alternative to our conflicted position. Refusing to participate in the world constructed by moderns and postmoderns is, however, not enough. I argue for an engagement in society and nature that is productive.

Figure 8.1 constructs four possibilities for valuing the concepts place and technology. So far I have argued that the early projects of Le Corbusier and the recent projects of the new urbanists hold down either end of the modern axis. They share the same assumptions, but simply invert how they are valued. Similarly, the projects of Rem Koolhaas and a constellation of four positive positions including critical regionalism, regenerative architecture, sustainability, and what Catherine Slessor calls eco-tech hold down either end of the nonmodern axis. The reader can understand *eco-tech,* from its location in the table, as a position that ties the environmental concerns of sustainability to high technology, as in the later works of Sir Norman Foster, Nicholas Grimshaw, Richard Rogers, or Renzo Piano.[25]

The producers of Blueprint Farm aspired to make a place that was sustainable, or regenerative, but produced instead a place that was critical. An axiomatic assumption of this inquiry has been that practice should influence theory to the same degree that theory influences practice. It is thus appropriate that I conclude by allowing the ironic discoveries of practice at Blueprint Farm to inform the renovation of critical regionalism as a nonmodern proposal for regenerative architecture.

Kenneth Frampton has, of course, nominated other projects as exemplars of critical regionalism. The works of Mario Botta in Switzerland, Jorn Utzon in Denmark, Luis Barragán in Mexico, and Alvaro Siza Viera in Portugal are among the most prominent. Other critics have nominated an Australian, Glenn Murcutt, as the most prominent of critical regionalists. Without diminishing the poetic and tectonic achievements of these architects, or the insight of Frampton's criticism, I must agree with Fredric Jameson that Frampton's discussion of these projects is largely aesthetic in character. My investigation of Blueprint Farm has engaged two additional issues—the political and the ecological—that are, I believe, essential characteristics of any architecture that aspires to be regenerative. The demand for an overtly political program comes not only from Jameson, but also from the advocates of social ecology. These observers require that architecture be understood, not in the aesthetic terms of high culture, but in the social and material context of everyday life. The demand for regenerative architecture to engage the ecology of places comes from the ecologists. These observers require that architecture be understood as the transformation of nature. The limits of a purely aesthetic discourse, critical though it may be, are that it remains outside the social and biological conditions that describe normative practices. The value of Blueprint Farm as a case study is that it ties together the aesthetic concerns of Frampton, the political concerns of Jameson, and the ecological concerns of the ecologists.

As I said at the beginning of this chapter, generalizing the eight propositions that summarize the ironic conditions of Blueprint Farm will constitute my proposal for a regenerative architecture. In Chapter 1, I proposed an a priori definition of regenerative architecture that was derived from the asocial definition offered by John Lyle. These generalizations will be stated as practice-based *attitudes,* not as deductive propositions. The term "attitude" is used by Frampton to describe the common features found in his own list of exemplars.[26] The distinction to be made between Frampton's list of seven points and my own list of eight is that his are *descriptive* and mine are *prescriptive.* Where Frampton has provided numerous positive examples that make his critical regionalism hypothesis concrete, I have provided only one example, that of Blueprint Farm, that ironically points the way toward a regenerative architecture. My assumption is, however, that we learn more from irony and/ or tragedy than from success.

EIGHT POINTS

FOR REGENERATIVE

ARCHITECTURE:

A Nonmodern

Manifesto

1. A regenerative architecture will construct social settings that can be lived differently.

This point rejects the notion that technology in itself might be an autonomous agent capable of liberating humans from the oppressive natural and/

or social conditions of place. Rather, it suggests that human institutions are both affected by and, in turn, affect the social construction of technological networks. Humans might, then, rationally and democratically construct regenerative technologies as the engaged agents of the humans and nonhumans that collectively inhabit a place.

2. So as to participate in local constellations of ideas, a regenerative architecture will participate in the tectonic history of a place.

Participation in the tectonic history of a place requires that the interventions of architects be, first, intelligible and, second, perceived as relevant to the material conditions of everyday life. These criteria in turn require that architectural projects, to the degree possible, relate to the three dimensions of technology: local *knowledge* and *practices,* and the *sets of objects* that contribute to a sense of place.

3. Rather than construct objects, the producers of regenerative architecture will participate in the construction of integrated cultural and ecological processes.

Historically, architects have tended to fetishize places as their own constructions and thus obscure the complex social and ecological processes that constitute architectural production. In lieu of understanding the desire of the communities we serve to live differently, architects have tragically obscured *intentional objects* by claiming sole authorship of *actual objects.* A regenerative architecture will deemphasize the significance of the actual object and emphasize the construction of processes that relate social activity to ecological conditions.

4. A regenerative architecture will resist the centers of calculation by magnifying local labor and ecological variables.

The overt political program of regenerative architecture will include two principal strategies: First, the producers of regenerative architecture will consciously subvert the universalizing and optimizing measures of objective building performance. These are typically promoted by such technological networks as the air-conditioning industry and measured in BTUs, calories, and watts. This strategy should not be construed to mean that human comfort is to be devalued. Second, regenerative architecture will rely upon *transparent* technologies to magnify local labor knowledge and local ecological conditions.

5. *Rather than participate in the aestheticized politics implicit in technological displays, regenerative architecture will construct the technologies of everyday life through democratic means.*

The market has increasingly manipulated architectural technology in order to stimulate those consumers who have become weary of the ever-increasing rates of production and consumption. A regenerative architecture will subvert the power of market-driven, sublime technology by engaging citizens in decision making about the technologies that enable everyday life. This strategy is recuperative in that it hopes to redescribe the meaning of *sublimity* as something closer to the ethical concept understood by Kant.

6. *The technological interventions of regenerative architecture will contribute to the normalization of critical practices.*

Rather than construct critical objects that inform viewers of how history might have been different, regenerative architecture will strive to influence normative construction practices. This proposition recognizes that the ontological dimension of building takes precedence over the representational—that the repetitive material practices of construction do more to influence the operation of society than do singular aesthetic critiques. In this sense, the reproduction of life-enhancing practices is preferred over aesthetic commentary.

7. *The practice of regenerative architecture will enable places by fostering convergent human agreements.*

A durable architecture need only delay the inevitability of decay. A sustainable architecture need only maintain the status quo of natural carrying capacity. A regenerative architecture, however, must concern itself with the reproduction of the institutional agreements that tie humans to the ecological conditions of a place. This suggests that architecture itself must facilitate democratic consideration of the tidal cycle, of prevailing breezes, or of the coolth of the earth itself. This is a matter of democracy and technological development.

   8. *A regenerative architecture will prefer the development of life-enhancing practices to the creation of critical and historically instructive places.*
The *critical place* helps society to understand that the social construction of places and technologies might have been different. Such a place is a memorial to the forgotten or as yet untried modes of noncapitalist production that would transform nature in some other way. My final point is that critical places are not in themselves productive. Better yet, *a critical place can become regenerative only through the production and reproduction of democratic, life-enhancing practices.*

THE THINGS THEMSELVES

In Chapter 5, "Technological Interventions," I analyzed wind-towers as one example of the nonconventional technologies employed by the ecologists at Blueprint Farm. In that context I argued, after Sal Restivo, that the development of these nonconventional technologies amounted to "a program for freedom and liberty in everyday life."[1] To make this argument comprehensive it would be necessary for me to subject each of the so-called nontraditional technologies proposed by either the Israeli engineers or the Texan ecologists to the same level of social and technical analysis to which I subjected wind-towers. That project, I'm afraid, would exhaust the patience of both the writer and the reader—it is a project best left to another publication. Rather than leave the reader simply guessing, however, I wrote this appendix to provide an abbreviated description of each nonconventional technology incorporated into the project.

My definition of "nonconventional" technology is a rather simple one—it means two things: first, that the technique of construction is uncommon, and second, that the technique was consciously intended to realize the goal of sustainability. By these criteria I have identified fifteen nonconventional technologies employed at the farm. Of these, I have already discussed wind-towers at some length and Icono-metrics™ to a lesser extent, leaving thirteen others to be described in this appendix. In each case I will briefly review, in no particular order, how the technology worked (or was proposed to work), its intended purpose, who developed it, and whether the technology has been reproduced elsewhere. Further conclusions are left to the reader.

STRAW-BALE WALLS   Non-load-bearing exterior walls of the office and packing sheds were constructed of straw or buffle grass bales laid in a masonry-like pattern. Walls were reinforced with wood at window and door openings and reinforced vertically with steel rods. The straw bales were then finished with spray-applied fly-ash stucco. Straw is a waste product of wheat farming—this was the material used in the first set of buildings. Buffle grass is a local, but invasive, species of pasture grass—this was the material used in the second set of buildings. Fly-ash is a waste by-product of coal burning and, when mixed with cement, makes a high-quality stucco.

The builders of Blueprint Farm articulated three reasons for building the walls of the packing sheds in this manner:

1. To provide high thermal insulation values with low embodied energy,
2. To harvest, and thus "repair," the invasion of non-native grass species in the region, and
3. To recycle waste products from coal burning.

The Center for Maximum Potential Building Systems was responsible for the engineering of the straw-bale walls constructed at the site. This technology has, however, a long history in both Europe and the United States that precedes the construction of Blueprint Farm. The reemergence of straw-bale building in the United States has been investigated by the sociologist Kathryn Henderson through the lens of science and technology studies.[2] Alternative

Figure A.1. Workmen assembling a straw-bale wall, photograph by Pliny Fisk III, © 1989 CMPBS. Courtesy of the Center for Maximum Potential Building Systems.

press books, such as *The Straw Bale House,* document the phenomenon (and the role of CMPBS in it) in a less critical light.[3] Although CMPBS is not solely responsible for the reemergence of straw-bale technology in the United States, it can certainly be argued that the visibility of Blueprint Farm contributed to the increasing popularity of the technology.

The refrigeration system proposed for the packing sheds at Blueprint Farm was an *absorption*-type system, rather than the more conventional *compression*-type system found in most U.S. commercial applications. Compression-

SOLAR/ZEOLITE REFRIGERATION

type systems operate by compressing, and thus raising, the temperature of a refrigerant liquid. The heat of compression is then wasted to the exterior environment by passing it through a decompression valve. The now decompressed, cooled liquid refrigerant is then passed through a heat exchanger coil to cool the space in question. Ecologists object to standard compression-type refrigeration because most of the liquid refrigerants used are environmentally hazardous, the compressors used to drive the systems are large consumers of electrical energy, and those same compressors require constant maintenance.

In contrast, absorption-type refrigerators are mechanically simple devices that operate by using a desiccant—a water-absorbing agent, in this case the mineral zeolite—to extract heat from the space in question via a water-based cycle that depends upon natural energy flows, rather than electricity, to boil and freeze the zeolite. With far fewer moving parts than conventional compression-type refrigerators, and free solar energy abundantly available in Laredo, the life cycle of absorption-type refrigerators is significantly greater. Rather than wasting heat to the ambient environment, the system was designed to use waste heat to produce hot water used elsewhere on site.

In the words of Howard Reichmuth, designer of the system for CMPBS:

The solar-zeolite refrigeration designed here is intended to cool a storage room, approximately a 10 foot cube. The principle of solar zeolite refrigeration is simple: In a permanently sealed system at near vacuum conditions, the zeolite (a reasonably common light brown mineral) is heated by sunlight during the day. The heat drives absorbed water out of the zeolite whereupon it gives up its heat in a radiator where it condenses in a container in an ice bath. The following night, the zeolite cools and aggressively demands its water back. The water in the container is evaporated at the near vacuum conditions causing it to freeze. The proposed solar zeolite refrigerator builds on prior successful demonstrations of the principles, but would have improved performance because concentrated sunlight is used to increase the temperature to which the zeolite is heated.[4]

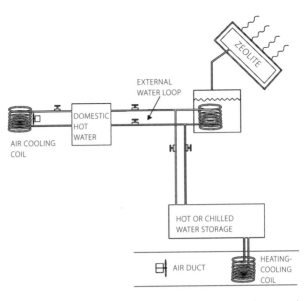

EXTERNAL WATER LOOP

ZEOLITE

DOMESTIC HOT WATER

AIR COOLING COIL

HOT OR CHILLED WATER STORAGE

AIR DUCT

HEATING-COOLING COIL

Figure A.2. Proposal for a solar/zeolite refrigeration system by Howard Reichmuth of Hood River, Oregon, for CMPBS. Courtesy Howard Reichmuth.

According to Fisk, the intention in using nonconventional absorption-type refrigeration was threefold:

1. To provide a climate-controlled environment to store and process crops, thus reducing spoilage before shipping,
2. To create a market for zeolite, a waste by-product of coal mining, and
3. To reduce the environmental hazards associated with conventional compression-type refrigerants.[5]

The solar/zeolite refrigeration system proposed for Blueprint Farm was designed by Howard Reichmuth of Hood River, Oregon, as a consultant to CMPBS.[6] It was, however, never constructed because its estimated cost ($8,000) was, at the time, more than three times that of the conventional compression-type system. Unfortunately, by rejecting the high initial cost of the system, those who held the project's purse strings ignored the fact that the system's annual energy benefit ($900) would pay for itself in less than ten years.

In "How the Refrigerator Got Its Hum," Ruth Schwartz Cowan provides a compelling social history of both the conventional compression-type refrigeration system and a nonconventional absorption system that is salient to this case (see Chapter 5).[7] Although I know of no other attempts to construct solar/absorption refrigeration systems in South Texas, such systems are becoming increasingly common in Europe and those locations where long-term economic thinking prevails. Thus, Blueprint Farm may have contributed to the increased visibility of this nonconventional technology.

SOLAR FOOD DRYER   The solar food dryer at Blueprint Farm was conceived in a variety of configurations by CMPBS and its consultants. The most feasible was that designed by engineer Howard Reichmuth for CMPBS. He described the design as

a semi-industrial scale food dryer capable of drying more than 1000 pounds of food a day. The dryer uses a large but inexpensive 100 foot by 16 foot inflated solar collector to heat 4000 cfm air flow by 20–30 degrees Fahrenheit. The key to the efficiency of the system is the use of a counter-flow drying arrangement where the hottest and driest air is passed over the driest food first. This counter-flow process is achieved by placing the food to be dried on wheeled racks which are rearranged every day during the four-day drying cycle. The economics of solar crop drying require a very low cost solar collector because the whole process can only be used for a month in a year.[8]

There was a single purpose in constructing the food dryer at the farm: By drying tomatoes, for example, rather than selling them in bulk for sauce, produce could be marketed at dramatically higher prices. The food dryer, then,

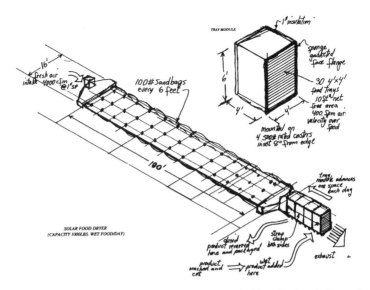

TRAY MODULE

1" insulation
sponge gasketed space flange
30 4'x4' food trays
105ft net free area
400 fpm air velocity over food
mounted on 4 speed rated casters inset 8" from edge

16'
fresh air intake 4000 cfm @ 1'sp
100# sandbags every 6 feet

100'

tray module advances one space each day

SOLAR FOOD DRYER
(CAPACITY 1000LBS. WET FOOD/DAY)

dried product removed here and packaged
strap clamp both sides
exhaust
product washed and cut → product added here wet

Figure A.3. Sketch of the solar food dryer proposed for Blueprint Farm by Howard Reichmuth of Hood River, Oregon, for CMPBS. Courtesy Howard Reichmuth.

was central to the value-added market strategy proposed by the Hightower regime.

Unfortunately, there is no documentation of the operation of the system designed by Reichmuth. Verbal accounts claim that, although CMPBS constructed the dryer as designed, the Israeli farm manager never used it.

The solar drying of foods is hardly a new food preservation technique. In the context of South Texas agriculture, however, the system was certainly unconventional. I do not know of any locations in South Texas where this technology has been reproduced.

SOLID-WASTE COMPOSTER

This technology was an in situ system for converting organic wastes gathered from Laredo Junior College cafeterias and local restaurants into crop fertilizer. Solid, organic wastes were arranged in windrows within the same pole structure bay to be planted in the next growing season. Windrows were covered in polyethylene tentlike channels connected to a reversible fan. The fan

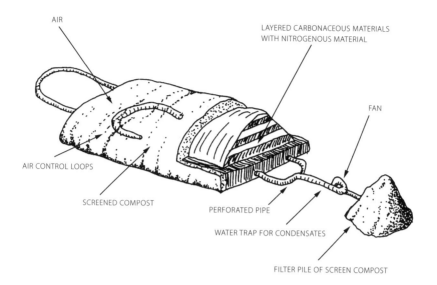

AIR

LAYERED CARBONACEOUS MATERIALS
WITH NITROGENOUS MATERIAL

FAN

AIR CONTROL LOOPS

SCREENED COMPOST

PERFORATED PIPE

WATER TRAP FOR CONDENSATES

FILTER PILE OF SCREEN COMPOST

Figure A.4. Diagram of the solid-waste composting system proposed for Blueprint Farm. Redrawn from design by the U.S. National Park Service.

could supply the heat of decay to the packing sheds, greenhouses, or crop area, or exhaust gases in the opposite direction to ambient air. When the composting cycle was complete, the compost was plowed into the ground.

The intent of solid-waste composting on site was fourfold:

1. To provide high-quality fertilizer to the farm,
2. To reduce the volume of solid waste at the local landfill,
3. To prepare compost at the point of use and thus avoid the cost of transporting fertilizer to the site, and
4. To provide an inexpensive source of carbon-dioxide-rich air to the greenhouses.

Although Pliny Fisk III's father was an expert in the design of composting systems, the proposal for Blueprint Farm was derived from designs by the U.S.

National Park Service. Composting is certainly a traditional, if not conventional, method of waste disposal. What made Fisk's use of the NPS system so unconventional was his proposal to construct the system in the location where fertilizer was needed and harvest the heat and carbon dioxide generated by organic activity. This technology has not, to my knowledge, been reproduced elsewhere in South Texas.

The composting toilet installed at Blueprint Farm was manufactured by Sunmar Ltd. of Canada, and was commercially available at the time of construction. The unit was not, however, approved by the local plumbing code. Because the installation was on state-owned land, local code-compliance was not required. The toilet composts solid and liquid human waste into an organic, nontoxic humus.

COMPOSTING TOILETS

The stated purpose for introducing this technology was to reduce water consumption on site. Recent revisions to state and local plumbing codes make the use of such systems more common, if not entirely conventional.

Figure A.5. The composting toilet installed at Blueprint Farm, photograph by Pliny Fisk III, © 1989 CMPBS. Courtesy of the Center for Maximum Potential Building Systems.

FERTIGATION™, OR
DRIP IRRIGATION

Fertigation™ is the proprietary name for an Israeli-engineered system of pumps, subterranean irrigation piping, and valves that distributes water to agricultural plots. The manufacturer claims that "The system can be operated by computer, enabling maximum control with minimum labor as well as individual fertigation regimes for each plot."[9]

The Israeli manufacturer stated three purposes for the use of this technology:

1. To conserve water by supply directly to crop roots, thereby minimizing evaporation and surface runoff losses,

2. To provide a liquid medium for the supply of fertilizers, fungicides, and fumigants, and

3. To increase the productivity of small plots of land.

The system installed at Blueprint Farm was developed by Netafim, Ltd., and engineered by Tahal Consulting Engineers, Ltd., Tel Aviv, Israel. The initial research and development of commercially scaled drip irrigation systems was

Figure A.6. Fertigation™, above-ground valves and piping. Author's photograph.

conducted by Texas A&M University and the U.S. Agricultural Extension Service at Lubbock, Texas. This basic design was developed as a commercially viable product by Netafim, Ltd. By 1995 a few medium-sized commercial farms in the Lower Rio Grande Valley had adopted the Israeli Netafim system.

INORGANIC MULCH

This technology consists of industrially produced polyethylene sheets that are spread over the soil during planting. Small holes are punched in the plastic to enable the insertion of seeds or seedlings into the soil. The system requires about 160 pounds of .003 mm plastic sheet per acre and mechanized tractor-based equipment. The Israeli engineers claimed that "Plastic mulch is effective in raising soil temperature.... By preventing direct contact between the plant (and fruit) and the wet soil, plastic mulch reduces spoilage by rotting. Another advantage of this method is earlier ripening of certain [cucurbits] and solanaceous species by two or three weeks in the spring," thus gaining a market advantage.[10]

This technique, first derived from the methods of French intensive gardening that became popular in the United States in the 1960's, was then adapted

Figure A.7. Inorganic mulch. Author's photograph.

Figure A.8. Israeli greenhouses in deteriorated condition. Author's photograph.

to a more industrial scale of agriculture by Tahal Consulting Engineers, Ltd., of Tel Aviv. By 1995 a small number of large farms in the Rio Grande Valley had adopted the system. A market has been found in Mexico for recycling the plastic sheeting, which is considered a waste product by American farmers.

GREENHOUSES   The initial recommendation of Tahal Engineering Consultants was for "plastic covered high tunnels" because of their low cost and ease of construction from local materials. This engineering recommendation was, however, passed over in favor of a more sophisticated, and expensive, greenhouse system fabricated with steel frames and a polyethylene cover manufactured in Israel.

The greenhouse technology realized three intentions:

1. To "protect the plants from low temperature, rain, hail, and winds; ...
2. "the large volume of air surrounding the plants facilitates ventilation and makes for better control of excessive heat and humidity." And,
3. "Certain plants ... require artificial illumination in order to regulate plant growth."[11]

The terrarium and the greenhouse have a long history of use in Europe and the United States. The technology can be considered nonconventional only by the scale of its use at Blueprint Farm. No other examples of Israeli-manufactured greenhouses are known in South Texas.

The shade structure at Blueprint Farm consisted of a **SHADE STRUCTURE** canopy of polyethylene netting stretched over a 30′ x 30′ grid constructed of recycled oil pipe stems used as columns and braced with steel cable. The shade cloth was, at least in theory, of various opacities, depending upon the degree of shade required.

Four intentions for this technology were made explicit by the Israeli engineers and the Texan ecologists:

Figure A.9. Shade structure in deteriorated condition. Author's photograph.

1. "This type of shade prevents direct damage to plants by hail and mitigates losses caused by strong wind and sandstorms, which not only injure the plants, but expose them to infection by disease causing organisms.

2. "There is also a moderating effect on extreme temperatures [in both summer and winter], . . .

3. "The 'roof' slows the movement of air, causing increased humidity. . . . there are plants, such as some flowers and green ornamentals, that thrive best in a humid atmosphere. To some extent the humidity can be controlled by unfolding the side and top screens."[12] And,

4. To recycle the waste products from more conventional technologies that are used in the manufacture of the shade cloth itself.

Both the Israeli engineers and the Texan ecologists lay claim to the recommendation to use shade structures at Blueprint Farm. The original June 1987 evaluation and planning study by Tahal Consulting Engineers proposes the use of "screen houses" and "windbreaks" that are described in terms very similar to the subsequent CMPBS design.[13] The Israelis refer to Middle Eastern tents as precedents, while the Texan ecologists refer to the traditional Native American "ramada" as the inspiration. In purely practical terms, local ranchers typically recycle oil pipes from closed wells as fence posts, and the steel cable used to stabilize the structure was appropriated from telephone pole rigging.

The system installed at the farm was engineered by CMPBS based upon the recommendations of the netting manufacturer. Unfortunately, the manufacturer failed to recommend that fabric panels be isolated from wave harmonic motion that is induced by wind. This structural flaw led to the premature deterioration of the canopy. No reproductions of the system are known.

**SOLAR
WATER HEATING**

The solar water heater fabricated at Blueprint Farm was a simple *batch-type* heater that operated without pumps, valves, or any sophisticated system of selective coatings or storage. Rather, a

Figure A.10. Solar water batch-heater installed at Blueprint Farm. Author's photograph.

vessel fabricated from off-the-shelf stainless steel cylinders, two salad bowls, and a pizza pan simply absorbed available insolation, which increased the water temperature in the reservoir to whatever temperature could be reached during daylight hours. When farm staff turned on the tap, hot water flowed by gravity, and a float valve allowed the reservoir to be refilled.

The rationale behind this simple system was to provide the minimum amount of hot water required to meet incidental staff needs at minimum cost. This design was previously developed by CMPBS and deployed at the Crystal City, Texas, *colonia*. Batch-type solar heaters are a common technology in rural Mexico. This system is nonconventional in that it employs common components toward uncommon ends. Approximately five hundred units of the CMPBS design are in use in the border region.[14]

SETTLEMENT POND

Water taken from the Rio Grande is not only very silty, but laden with multiple pollutants, including heavy metals. The primary settlement system was a polyvinyl-lined excavation in the earth sized to support irrigation demand. A sec-

Figure A.11. The settlement pond at Blueprint Farm in deteriorated condition. Author's photograph.

ondary system of vegetative filters was proposed by CMPBS, but never funded or installed.

The primary purpose of the pond was to allow silt and other solids to settle out of irrigation water before it was pumped into the fields via the Fertigation™ system engineered by Tahal Engineering Consultants. The pond was to be periodically drained so that the polluted sludge could be extracted. The initial feasibility study by Tahal recommended the pond's construction and calculated the required capacity.

It was CMPBS, however, that recommended the secondary vegetative filtering system based upon research by John Todd, the Canadian-born marine biologist who is now director of Living Technologies of Burlington, Vermont. The idea of vegetative filters goes back to the ancient Maya, who used floating beds of water hyacinth to filter their drinking water and as a source of organic composting material. Water hyacinth, and many other plants, consume those biological entities that are harmful to humans. Contemporary scientists like Todd have designed such vegetative filter systems for large municipal drinking water systems.[15] Agricultural settlement ponds are common in Texas, but none is known to include designed vegetative filters.

Each of the five tetrahedron roofs of the packing sheds at Blueprint Farm collected rainwater and conveyed it to ad- jacent cisterns. This stored water was used for miscella- neous staff needs, excluding drinking, and for the evapo- rative Hexel pads incorporated in the wind-tower cooling system.

The intent of CMPBS in harvesting rainwater was, first, to conserve water from city mains and reduce demand on the river and, second, to provide thermal mass as a heat sink available to the solar/zeolite refrigeration system (although this potential is not mentioned by Howard Reichmuth, who engineered that system). The rainwater system was designed by CMPBS. In the years since the construction of Blueprint Farm, rainwater catchment systems have become increasingly common in Texas—around Austin in particular. There is no doubt that Blueprint Farm and CMPBS have played a significant role in the popularization of this technology. Cities all over the Southwest now include provisions for rainwater catchment systems in their municipal plumbing ordinances.

Figure A.12. Downspout from roof and cistern at Blueprint Farm. Author's photograph.

Figure A.13. Wind-powered turbine generator at Blueprint Farm. Author's photograph.

WIND-POWERED TURBINE GENERATORS

The four "windmills" installed at Blueprint Farm were commercially available units designed and manufactured by Burgey Wind Turbines, Inc., for CMPBS. Their purpose was to provide electricity to two submerged pumps in the Rio Grande intended to lift water to the settlement pond, and to two additional pumps intended to distribute water to the fields via the Fertigation™ system. Because these devices produced electricity, rather than powering pumps via mechanical means, they are better described as "wind turbines" than "windmills." Torque-type windmills are a common sight in the Texas landscape. They have been used to pump water since the era of settlement. The wind turbines installed at Blueprint Farm were not, however, of the traditional multibladed design. Several observers suggested that the traditional design would have worked far better because, in the traditional arrangement, the pump is located at the top of the water line, away from the silt and contaminants that caused constant maintenance problems for the pumps at Blueprint Farm.

NOTES

1. Heidegger's view toward nature is alternately claimed as support for those who adopt postmodern assumptions and those who adopt neo-Aristotelian assumptions. Such competing claims are not, in my view, mutually exclusive. Like Richard Rorty, I prefer to categorize Heidegger's ontology as postmodern because it is so clearly a rejection of the modern Cartesian dualism. It is less clear to me that Heidegger intended a recuperation of Aristotle.

2. "Dr. Héctor Jiménez," author interview, June 21, 1995.

3. "Alvaro Lacayo," author interview, June 23, 1995.

4. Throughout this text I will refer to Pliny Fisk III and his collaborators at Blueprint Farm as "ecologists" rather than "environmentalists." This political/philosophical distinction has been made by David Harvey, who states that "'environmentalists' . . . adopt an external and often managerial stance toward the *environment* and 'ecologists' . . . view human activities as embedded in *nature* (and . . . consequently construe the notion of human health in emotive, esthetic as well as instrumental terms)." See David Harvey, *Justice, Nature & the Geography of Difference,* p. 118.

5. See Anna Bramwell, *Ecology in the Twentieth Century: A History.*

6. See John Tillman Lyle, *Regenerative Design for Sustainable Development,* p. 10.

7. Harvey, *Justice, Nature & the Geography of Difference,* p. 148.

8. For a previous attempt to define this term in relation to the concept of "value," see Steven A. Moore, "Value and Regenerative Economy in Architecture," in *Proceedings of the ACSA Annual Meeting at Dallas, March 15–18, 1997,* pp. 544–551.

9. For a review of four titles on sustainable architecture published in the late 1990's that ignore the cultural content of technology, see Steven A. Moore, "Book Review Essay," *Journal of Architectural Education,* Spring 2000, pp. 245–249.

10. Peter Collins has made the same argument, and in greater detail. See *Changing Ideals in Modern Architecture 1750–1950.*

11. These two projects, human emancipation and self-realization, are described by David Harvey as the "twin ideals" of the Enlightenment. See Harvey, *Justice, Nature & the Geography of Difference,* p. 121.

12. Richard Rorty, the postmodern pragmatist, is a principal voice for those humble "ironists" who doubt their own "final vocabularies." Rorty reasons that "On my account of Ironist culture, . . . we should stop looking for a successor to Marxism, for a theory which fuses decency with sublimity. Ironists should reconcile themselves to a private-public split within their final vocabularies, to the fact that the resolution of doubts about one's final vocabulary has nothing in particular to do with attempts to save other people from pain and humiliation." For Rorty, the danger is in imagining that we can get things right. To act upon our right knowledge of history will invariably bring "pain and humiliation" in unforeseen ways. See *Contingency, Irony, and Solidarity,* pp. 90–95.

13. See John Agnew, *Place and Politics,* p. 62. Agnew also discusses the theme of the historic devaluation of place in "Representing Space: Space, Scale and Culture in Social Science."

14. For example, crime statistics document that the murder rate in New York City is significantly less than that of rural Arkansas. See Box Butterfield, "Nationwide Drop in Murders Is Reaching to Small Towns," *New York Times,* May 9, 2000.

15. Although Weber is commonly credited with the coinage of these terms, they belong to Ferdinand Tonnies, who first used them in 1887. See Tonnies, *Community and Society* (New York: Harper, 1963).

16. Agnew, *Place and Politics,* p. 231.

17. Soja's position is associated with the tradition of critical theory; however, his intention is revisionist. See Edward Soja, *Postmodern Geographies: The Reassertion of Space in Critical Social Theory,* p. 120.

18. Anna Bramwell, for example, has argued that German anti-Semitism arises from the doctrines of environmental determinism. To generalize that all Germans share a genius that originates in the forest and that wandering Jews share a rootlessness that originates in the desert is a classic example of determinist reductivist logic. See Anna Bramwell, *Blood and Soil: Richard Walter Darre and Hitler's Green Party.* See also Jeffery Herf, *Reactionary Modernism: Technology, Culture and Politics in Weimar and the Third Reich.*

19. David Harvey, *The Condition of Postmodernity,* p. 273. Harvey later softens his position with regard to the concept of place. In *Justice, Nature & the Geography of Difference,* he accepts Raymond Williams's proposal for "militant particularism," in which local practices are seen as resistant to the tendency of flexible capital to appropriate locally produced value. This more recent position is sympathetic to Frampton's critical regionalism hypothesis.

20. R. J. Johnston argues that territory is defined by both inclusive and exclusive operations. "Inclusion" creates territory through enforcement of the "law of the land." "Exclusion" operates through practices such as zoning or overt segregation. In both operations, the world is segmented, leading to inevitable territorial conflict. See R. J. Johnston, *A Question of Place*, p. 210.

21. Le Corbusier, "Guiding Principles of Town Planning," p. 91.

22. Alexander Tzonis and Liane Lefaivre, "Critical Regionalism," p. 8.

23. Jonathan Smith has suggested that the antimodern, antitheoretical trend of contemporary architecture may simply be the expression of market forces upon the discipline. In this view, clients tell architects what to build, they don't ask them. While such market pressure certainly guides most practitioners, it does not account for the position of those who, like Leon Krier and his mentor Prince Charles, are economically insulated from such pressures.

24. The term "critical regionalism" was first used by Alexander Tzonis and Liane Lefaivre in "The Grid and the Pathway." These authors have contributed to the development of critical regionalism as a highly regarded polemic for architectural production. Kenneth Frampton, however, has been regarded as the principal author of the position in a series of essays published between 1983 and 1991. These include: "Toward a Critical Regionalism: Six Points for an Architecture of Resistance" (1983); "Prospects for a Critical Regionalism" (1983); "Architecture and Critical Regionalism" (1983); "Critical Regionalism: Modern Architecture and Cultural Identity" (1985); "Place-form and Cultural Identity" (1988); "Critical Regionalism Revisited" (1991).

25. I am indebted to Karen Cordes Spence for this insight. She argues that each of Frampton's seven points is the embodiment of a dialectic and refers to Frampton's foundation in critical theory and the modern subject/object split. See her doctoral dissertation, "Theorizing in Recent Architecture: An Examination of the Texts of Frampton, Rossi, and Lang."

26. Frampton, "Toward a Critical Regionalism," pp. 16–30. Frampton's position would benefit by considering that of Deleuze and Guattari, who argue that "it would be wrong to confuse isomorphy with homogeneity. For one thing, isomorphy allows, and even incites, a greater heterogeneity among states (democratic, totalitarian, and, especially, 'socialist' states are not facades).... the internationalist capitalist axiomatic effectively assures the isomorphy of the diverse formations only where the domestic market is developing and expanding.... But, it tolerates, requires ... a certain peripheral polymorphy, to the extent that it is not saturated, to the extent that it repels its own limits." By insisting upon the tendency of capital to homogenize place, Frampton fails to recognize that isomorphic capital may find the heterogeneity of places to be a market opportunity for the extraction of value. Although this possibility is not foreseen by Frampton, it does not, I think, subvert his argument in support of local places as resistant to the hegemonic forces of capital. See Gilles Deleuze and Felix Guattari, *A Thousand Plateaus: Capitalism and Schizophrenia,* p. 436.

27. Frampton, "Critical Regionalism: Modern Architecture and Cultural Identity," p. 327.

28. Frampton states,"Like many others of my generation I have been influenced by a Marxist interpretation of history." See Frampton, *Modern Architecture: A Critical History*, p. 9.

29. Theodor W. Adorno,"Fetish Character in Music and Regression of Listening," p. 298. Although Frampton departs from Adorno's negativist tactics in many regards, the two share an abiding suspicion of populist sentiments. It is the elitist, or formalist, position that Frampton shares with Adorno that made him an early and vehement critic of the postmodern populism advanced by Robert Venturi in the 1960's.

30. For their part, Tzonis and Lefaivre have recommended that critical regionalism engage in the tactic of "defamiliarization," where architecture demands of the viewer a thoughtful consideration of how the building operates in its particular setting, free from comforting historical references. This tactic is entirely consistent with Frampton's position and a generally modernist depth model advanced by critical theory. See Alexander Tzonis and Liane Lefaivre,"Critical Regionalism," pp. 2–13.

31. For a discussion of Adorno's critical aesthetic theory, see David Held, *Introduction to Critical Theory: Horkheimer to Habermas*.

32. Bookchin has been a visible activist since the 1950's. His most well-known titles on the topic of social ecology include: *The Ecology of Freedom* (1991), *The Limits of the City* (1973), *The Modern Crisis* (1986), *The Philosophy of Social Ecology* (1995), *Post-Scarcity Anarchism* (1971), and *Toward an Ecological Society* (1980). He is also editor of the journal *Society and Nature*.

33. Harvey, *Justice, Nature & the Geography of Difference*, p. 48.

34. See Richard Ingersoll,"Second Nature: On the Bond of Ecology and Architecture." See also Andrew Feenberg, *Critical Theory of Technology*, pp. 175–177.

35. Frampton,"Seven Points for the Millennium: An Untimely Manifesto," *Architectural Review* 206, no. 1233 (1999): 78, 79, 80. The same project of renovation is evident in his "Themes and Variations," lecture delivered at the University of California, College of Environmental Design, January 27, 2000.

36. This definition has been adopted by the International Union for the Conservation of Nature (IUCN). See William McDonough, *The Hannover Principles*. Many observers find this definition to be less anthropocentric, and thus more acceptable, than the more popular definition, which reads,"improving the quality of life without detracting from the resources available to future generations."

37. This definition of "carrying capacity" is from Wendell Berry, *The Unsettling of America: Culture and Agriculture*, p. 7.

38. The doctrines of deep ecology are most clearly stated by the Norwegian philosopher Arne Naess. Central to Naess's position is the notion that the self is indistinguishable from the natural world, as opposed to the independent self identified by Cartesian philosophy. The interdependent self, as understood by deep ecology, proclaims the intrinsic value of other life forms and the dependence of the "self-as-part-of-nature" upon those other life forms. See *Ecology, Community, and Lifestyle*.

39. See Fredric Jameson, *The Seeds of Time*, pp. 187–203. Jameson provides a

very thoughtful critique of Frampton's proposal for critical regionalism as the conclusion to this essay.

40. Frampton, "Themes and Variations."

41. Jameson, *The Seeds of Time,* p. 46. Jameson's text more completely reads as: "But surely ecology is another matter entirely; and while its rediscovery and reaffirmation of the limits of nature is postmodern to the degree to which it repudiates the modernism of modernization and of the productivist ethos that accompanies an earlier moment of capitalism, it must also equally refuse the implied Promethianism of any conception of Nature itself, the Other of human history, as somehow humanly constructed. How antifoundationalism can thus coexist with the passionate ecological revival of a sense of nature is the essential mystery at the heart of what I take to be a fundamental antinomy of the postmodern."

42. Ibid., pp. 190–193. Jameson's intention is, of course, to suggest a possible path for the renovation of critical regionalism that would be consistent with his own foundationalism.

43. See Kenneth Frampton, *Studies in Tectonic Culture.* Semper's insight into the anthropological origins of the tectonic is also investigated by Frampton in "Rappel à l'ordre."

44. Frampton, *Studies in Tectonic Culture,* p. 2.

45. For a definition of "technological networks" by Latour, see *We Have Never Been Modern,* p. 118. My interpretation of the philosophical concept of "intentionality" is largely derived from that of Hubert Dreyfus in *Husserl, Intentionality and Cognitive Science.*

Chapter Two

1. The text of Jackson's remarks is as follows: "Jewtown is where Hymie gets you if you can't negotiate them suits down." This is cited in Leonard Dinnerstein, *Anti-Semitism in America,* p. 219. Dinnerstein makes it clear that this text was neither taken out of context nor inconsistent with Jackson's attitudes expressed on other occasions.

2. Texas Department of Agriculture, "The Texas-Israel Exchange."

3. Texas Department of Agriculture, "The Dissemination of Innovation: Training and Educational Development on the 'Blueprint' Demonstration Farm, Laredo, Texas. A Proposal Submitted to the Hitachi Foundation October, 1998," p. 2.

4. "Alvaro Lacayo," author interview, June 23, 1995.

5. To be clear, Fisk did not study in the Louis Kahn studio at Penn. The influence of Kahn upon his work, however, is distinct. Fisk did study with Ian McHarg at Penn and has maintained a relationship with him in the intervening years.

6. Pliny Fisk III, memo to Gideon Yogev, Jack Stallings, and Nancy Epstein, April 25, 1988, CMPBS File 3.4, p. 3. Photocopy available at the author's office, University of Texas at Austin.

7. Pliny Fisk III, memo to Nancy Epstein and John Volcek, September 22, 1988, CMPBS File 3.7, p. 1. Photocopy available at the author's office, University of Texas at Austin.

8. Texas Department of Agriculture, "Mission Statement for the Laredo Blueprint Demonstration Farm."

9. "Alvaro Lacayo," author interview, June 23, 1995.

10. Pliny Fisk III, author interview, February 29, 1996.

11. Texas Department of Agriculture, "Internal TDA Assessment of the Laredo Situation."

12. Pliny Fisk III, "General Issues Regarding Farm from CMPBS, June 1989," CMPBS File 3.8. Photocopy available at the author's office, University of Texas at Austin. This document was never made public. It is best characterized as an internal CMPBS memo.

13. Dale Burnett, Director, Pesticide Enforcement, Texas Department of Agriculture, to Pliny Fisk, CMPBS, March 20, 1990, CMPBS File 3.1. Documentation available at the author's office, University of Texas at Austin.

14. Pliny Fisk III to John Volcek, December 28, 1990, CMPBS File 3.9. Photocopy available at the author's office, University of Texas at Austin.

15. Nancy Epstein [TDA] to Pliny Fisk III, July 13, 1990, CMPBS File 1.9. Documentation available at the author's office, University of Texas at Austin.

16. Henri Lefebvre, *The Production of Space,* pp. 70–75.

17. Pliny Fisk III, author interview, February 26, 1996.

18. Donald A. Hegwood, Dean, to Mark Ellison, Assistant Commissioner for Marketing and Agricultural Development, January 7, 1991, Texas Department of Agriculture File (TDAF) 11.6. Photocopy available at the author's office, University of Texas at Austin.

19. Gregg H. Goldstein, Acting TIE Director, to Ms. Felicia B. Lynch, Senior Vice President, Hitachi Foundation, May 31, 1991, Texas Department of Agriculture File (TDAF) 11.2. Photocopy available at the author's office, University of Texas at Austin.

20. Nancy Epstein, "Memo to Mark Ellison Regarding the Texas-Israel Exchange: Background, Issues and Recommendations," January 30, 1991, Texas Department of Agriculture File (TDAF) 12.2, p. 6. Photocopy available at the author's office, University of Texas at Austin. Part of this file was blacked out by the state attorney general and classified as sensitive personnel information.

21. María Eugenia (Meg) Guerra, Project Director, to Pliny Fisk III, September 4, 1994, CMPBS File 1.28. Photocopy available at the author's office, University of Texas at Austin.

Chapter Three

1. Henri Lefebvre, *The Production of Space,* p. 83.

2. Ibid., p. 26.

3. See Neil Leach, *Rethinking Architecture: A Reader in Cultural Theory,* pp. 139–148.

4. Within social science, three traditions have evolved within science and technology studies: the network theory, pioneered by Bruno Latour and Michel Callon; the systems theory, developed by Thomas Hughes; and the social constructivist theory, argued by Bijker, Law, MacKenzie, Wajcman, and others. Each of these posi-

tions has important contributions to make to this discussion; it is only the lack of space that prohibits their inclusion. These recent theories of technology take advantage of the work of Thomas Kuhn in the 1950's and 1960's, but depend most directly upon the pioneering research of the Frankfurt School beginning in the 1930's.

5. Latour, *We Have Never Been Modern,* p. 117.

6. See Martin Heidegger, *Being and Time.* Those who have expanded upon Heidegger include Edward Casey, *Getting Back into Place: Toward a Renewed Understanding of the Place-world,* and David Seamon, *A Geography of the Lifeworld: Movement, Rest, and Encounter.*

7. John Agnew, *Place and Politics: The Geographical Mediation of State and Society,* p. 28. The definition of these terms is further amplified in his essay "Representing Space: Space, Scale, and Culture in Social Science."

8. The middle scale of inquiry to which Agnew refers, halfway between the macro scale of science and the micro scale of poetry, is supported by Thomas Misa in "Retrieving Sociotechnical Change from Technological Determinism," p. 116. Misa argues that those who employ macro strategies of inquiry tend to arrive at technologically determined arguments, and those who employ micro strategies tend to arrive at socially determined arguments. A "middle-level" analysis tends to avoid the errors of either overdetermination or underdetermination.

9. See Thomas Misa, "Retrieving Sociotechnical Change from Technological Determinism."

10. Agnew, "Representing Space," p. 263.

11. Reductive, physicalist definitions of technology tend to be less sophisticated in their understanding of the social construction of artifacts. However, Bruce Bimber, in "Three Faces of Technological Determinism," develops a very scholarly, yet reductive, definition of technology as limited to apparatus. Bimber's project, however, leads to other ontological problems beyond the scope of this study.

12. See Donald MacKenzie and Judith Wajcman, *The Social Shaping of Technology,* pp. 3–4.

13. Latour, *We Have Never Been Modern,* p. 117.

14. See Bruno Latour's discussion of "immutable mobiles" in "Visualization and Cognition: Thinking with Eyes and Hands," p. 9; see also *We Have Never Been Modern,* pp. 117–118.

15. Lefebvre, *The Production of Space,* p. 190.

16. Ibid., p. 31.

17. Anthony Giddens is credited with developing the theory of structuration, which is an attempt to synthesize the seemingly opposed principles of voluntarism and determinism. He argues that humans are free to transform social structures, but are also products of those structures. My argument here, regarding the relationship of places and technologies, is drawn from the same logic. See also MacKenzie and Wajcman, *The Social Shaping of Technology,* p. 6.

18. Philip Brey has examined how "space-shaping technologies" have disembedded the contemporary phenomenon of place. Where Brey's study has

focused upon the role of "connectivity development" in transforming the experience of place, my own emphasis has been on what Brey terms "local development." See Philip Brey, "Space-Shaping Technologies and the Disembedding of Place," p. 242.

19. See Sister Natalie Walsh, "The Founding of Laredo and St. Augustine Church" (Master's thesis, University of Texas, 1935), p. 1.

20. For an in-depth discussion of the Law of the Indies and the Spanish colonial ordinance, which dictated the pattern of land settlement and distribution in the colonies, see John Stilgoe, *Common Landscapes of America: 1580 to 1845*, pp. 34–42.

21. "Thomas Rosas," author interview, June 23, 1995.

22. The myth of the decline of produce production in la Frontera Chica is supported by statistics. From 1984 to 1985, the area producing vegetables in Webb County fell from 4,104 acres to 2,975 acres, while the cattle population grew by 7,000 head. See Dennis Findlay, ed., *Texas County Statistics*, p. 259.

23. Jim Hightower, *Hard Tomatoes, Hard Times*, p. 30.

24. "The point is, however, that where the thematic opposition of heterogeneity and homogeneity is invoked, it can only be this brutal process that is the ultimate referent: the effects that result from the power of commerce and capitalism proper—which is to say sheer numbers as such, number now shorn and divested of its own magical heterogeneities and reduced to equivalencies—to seize upon a landscape and flatten it out, reorganize it into a grid of identical parcels, and expose it to the dynamic of a market that now organizes space in terms of an identical value. The development of capitalism then distributes the value most unevenly indeed, until at length, in its postmodern moment, sheer speculation, as something like the triumph of spirit over matter, the liberation of the form of value from any of its former concrete or earthy content, now reigns supreme and devastates the very cities and countrysides it created in the process of its own earlier development. But all such later forms of abstract violence and homogeneity derive from the initial parcelization, which translates the money form and the logic of commodity production for a market back onto space itself." Fredric Jameson, *The Seeds of Time*, p. 25.

25. "Roberto Elizondo," author interview, June 24, 1995.

26. Hightower, *Hard Tomatoes, Hard Times*, p. 31.

27. This argument by Hightower sounds rather like one put forward by Fredric Jameson: "But the very possibility of a new globalization (the expansion of capital beyond its earlier limits in its second, or 'imperialist,' stage) depended on an agricultural reorganization (sometimes called the green revolution owing to its technological and specifically chemical and biological innovations) that effectively made peasants over into farmworkers and great estates or latifundia (as well as village enclaves) over into agri-business. . . . the second, or modern, moment of capital—the stage of imperialism—retained an older precapitalist mode of production in agriculture and kept it intact, exploiting in tributary fashion, deriving capital by extensive labor, inhuman hours and conditions, from essentially

precapitalist relations. The new multinational stage of capital is then character-ized by the sweeping away of such enclaves and their utter assimilation into capi-tal itself, with its wage labor and working conditions: at this point agriculture—culturally distinctive and identified in the super-structure as the Other of nature—now becomes an industry like any other, and the peasants simply work-ers whose labor is classically commodified in terms of value equivalencies." Jameson, *The Seeds of Time,* pp. 26–27.

28. "Roberto Elizondo," author interview, June 24, 1995.

29. "Alvaro Lacayo," author interview, June 23, 1995.

30. Hightower, *Hard Tomatoes, Hard Times,* p. 21.

31. "Alvaro Lacayo," author interview, June 23, 1995.

32. In 1911, Dean Liberty Hyde Bailey of Cornell University promoted the doc-trine of Social Darwinism as it is applied to agricultural mechanization. More re-cent federal policy, established under the U.S. Department of Agriculture (the administration of Earl Butz, in particular), has amounted to the same proposition—only the fittest farmers deserve to survive. See Jim Hightower, *Hard Tomatoes, Hard Times,* p. 14.

33. See William Eno de Buys, *Enchantment and Exploitation: The Life and Hard Times of a New Mexico Mountain Range.*

34. Willard Cochrane, *The Development of American Agriculture,* pp. 389–390. Cochrane's theory of the "technological treadmill" has been widely accepted by critics of agriculture in the United States. However, few have recognized that Cochrane's thesis is nearly identical to that proposed by Marx in *Wage Labour and Capital* over a century ago. He argues that "The aggressive, innovative farmer is on a treadmill with regard to the adoption of improved technologies on his farm. As he rushes to adopt a new and improved technology when it first becomes avail-able, he at first reaps a gain. But, as others after him run to adopt the technology, the treadmill speeds up and grinds out an increasing supply of the product. The increased supply of the product drives the price of the product down to where the early adopter and his fellow adopters are back to a no-profit situation. Farm technological advance in a free market situation forces the participant to run on a treadmill. For laggards who never got on the treadmill, the consequences of farm technological advance are more devastating. Farm technological advance either forces them into a situation of economic loss or further widens their existing losses.

"As these laggard farmers have been forced out of business by the process of farm technological advance, their productive assets have typically been acquired by better, more aggressive farmers—by the 'early-bird' farmers who prospered by the temporary gains of the early adoption of the new and improved production technologies. . . . The process of farm technological advance has contributed im-portantly to the redistribution of productive assets in American Agriculture in which commercial production has been, and continues to be, concentrated on large farms."

35. Joel Garreau describes "Mex-America," the 1,933-mile border between Mexico and the United States, as one of the nine distinct regions of North America.

He includes in this culturally distinct region, not only the twin border towns that mark national territories, but the metropolitan centers such as San Antonio, Albuquerque, and Los Angeles that service the border economy. See Joel Garreau, *The Nine Nations of North America,* p. 211.

36. This term is borrowed from the Laredo architect Viviana Frank.

37. See Gloria Anzaldúa, *Borderlands La Frontera: The New Mestiza,* p. 20.

38. "Alvaro Lacayo," author interview, June 23, 1995.

39. "Celia Juárez," author interview, June 20, 1995.

40. "Alvaro Lacayo," author interview, June 23, 1995.

41. For a full examination of the social and economic phenomenon of the *maquiladora,* see Patricia Wilson, *Exports and Local Development: Mexico's New Maquiladora,* pp. 36–37, 114.

42. Rafael Longoria, "Disparity and Proximity," *CITE: The Architecture and Design Review of Houston* 30 (Spring–Summer 1993): 11.

43. De Buys, *Enchantment and Exploitation,* p. 297.

44. "Vera Sassoon," author interview, June 21, 1995.

45. Sissy Farenthold, "Commerce without Conscience," *CITE: The Architecture and Design Review of Houston* 30 (Spring–Summer 1993): 12–14.

46. "Alvaro Lacayo," author interview, June 23, 1995.

47. Fort McIntosh, named for an Anglo officer killed in the war with Mexico, was one of a series of forts built along the disputed Mexican American frontier immediately following the Treaty of Guadalupe Hidalgo, signed in 1848. Jerry Thompson, *Sabers on the Rio Grande,* p. 166.

48. Lefebvre, *The Production of Space,* p. 117.

49. Ibid., p. 151.

50. These are Spain, Mexico, the Confederate States of America, and the United States of America. To make matters more complex, three additional governments influenced conditions in Laredo. France, although it never claimed Laredo in the era of Maximilian, certainly exerted influence there. The same holds true for the Republic of Texas, which never claimed territory south of the Nueces River. Finally, the Republic of the Rio Grande, which existed only for one year, claimed Laredo as its capital, but was never officially recognized by other governments.

51. See Hermalinda Aguirre Murillo, *History of Webb County* (Master's thesis, Southwest State Teachers College, 1941), p. 6. The title "Gateway to Mexico" was given to Laredo upon the connection of the Texas-Mexican Railroad from Corpus Christi with the National Railway of Mexico at Laredo in 1881. That title will be magnified by the pending appropriation of federal funds to construct the NAFTA Highway from Laredo to Canada. See Dan Feldstein, "House Approval of NAFTA Highway Predicted," *Houston Chronicle,* August 30, 1995.

52. Walsh, *The Founding of Laredo and St. Augustine Church,* pp. 76–91. *Porciones* measuring 1,000 *veras* along the river and 30,000 *veras* in depth were apportioned by lottery.

53. Stilgoe, *Common Landscapes of America,* pp. 12–21.

54. De Buys, *Enchantment and Exploitation,* p. 175. De Buys and John Stilgoe

are in apparent disagreement on this point. Stilgoe argues that the settlers of Chimayo clung to the nucleated colonial grid and de Buys argues the opposite, that settlers dispersed after the threat of Indian attack disappeared. I will assume that both are correct. In other words, the settlers of Chimayo documented by Stilgoe may have retained the colonial pattern, but other settlers in the Sangre de Cristo Mountains documented by de Buys have dispersed. My own informal observations of that region support this resolution.

55. De Buys, *Enchantment and Exploitation,* p. 195.

56. See Lefebvre, *The Production of Space,* p. 116.

57. Ibid., p. 110.

58. This passage is cited in J. K. Gibson-Graham, *The End of Capitalism (As We Knew It): A Feminist Critique of Political Economy,* p. 26. It is attributed to Althusser, citing Marx, from *Politics and History,* p. 175.

Chapter Four

1. See William Morris, ed., *The American Heritage Dictionary of the English Language,* p. 1275.

2. See Dagfinn Follesdal, "Brentano and Husserl on Intentional Objects and Perception," pp. 34–35.

3. See Aron Gurwitsche, "Husserl's Theory of Intentionality and Consciousness," pp. 66–68.

4. Hubert L. Dreyfus, *Husserl, Intentionality and Cognitive Science,* p. 25.

5. See Gurwitsche, "Husserl's Theory of Intentionality and Consciousness," p. 65.

6. Hubert Dreyfus, *Being-in-the-World: A Commentary on Heidegger's Being and Time, Division 1,* pp. 60–88.

7. Dreyfus also argues that the concept of phenomenological reduction is central to Husserl's version of *intentionality.* By reduction, Husserl means a "special act of reflection, in which we turn our attention *away from* the object being referred to (and *away from* our psychological experience of being directed toward the object), and turn our attention *to* the act, more specifically to its intentional content, thus making our representation of the conditions of satisfaction of the intentional state our object." See Dreyfus, *Husserl, Intentionality, and Cognitive Science,* pp. 1–6.

8. For a discussion of quasi-objects and quasi-subjects, see Brian Massumi, "Which Came First? The Individual or Society? Which Is the Chicken and Which Is the Egg?: The Political Economy of Belonging and the Logic of Relation."

9. Jim Hightower, author interview, November 14, 1995.

10. Ibid.

11. "Thomas Rosas," author interview, June 23, 1995.

12. "Alvaro Lacayo," author interview, June 23, 1995.

13. Jim Hightower, author interview, November 22, 1994.

14. Aant Elzinga argues that "all modern political systems [have] an interest in scientific knowledge that may serve to legitimize political action." In his view, the concept of the "epistemic regime" begins to explain how there is "selectivity, direc-

tion, and strategic action in research." For Elzinga, a regime is "a set of institutional arrangements, or sets of implicit and explicit decision-making procedures," that control expectations. In his several case studies, Elzinga demonstrates that the distinction among scientific inquiry, administrative policy, and political action is, in practice, blurred. He argues that if controversies exist in these cases, they should be understood as "science-based controversies," not "scientific controversies."

The Hightower network was a "technological regime" in the sense intended by Elzinga. The controversy that took place in Laredo between the Hightower network and the ecologist network was a "science-based controversy." The confrontation was not over the determination of "sustainable" farming methods, but who would control the useful political capital of "sustainability" as a concept. See Aant Elzinga, "Science as the Continuation of Politics by Other Means."

15. "Alvaro Lacayo," author interview, June 23, 1995.

16. "Dr. Héctor Jiménez," author interview, June 21, 1995.

17. Ibid.

18. "Celia Juárez," author interview, June 20, 1995.

19. "Rafael Bernadini," author interview, June 20, 1995.

20. "Alvaro Lacayo," author interview, June 23, 1995.

21. Ibid.

22. "Tim Deere," author interview, February 8, 1996.

23. Pliny Fisk III, author interview, February 29, 1996.

24. "The time has come to integrate an ecological natural philosophy with an ecological social philosophy based on freedom and consciousness, a goal that has haunted western philosophy from the pre-Socratics onward. Doubtless, the practical implications of this goal are paramount. If we are to survive ecological catastrophe, we must decentralize, restore bioregional forms of production and food cultivation, diversify our technologies, scale them to human dimensions, and establish face to face forms of democracy." Murray Bookchin, *Toward an Ecological Society,* p. 27.

25. Pliny Fisk III, "Interview: Pliny the Greener," *Architecture,* June 1998, p. 55.

26. "Alvaro Lacayo," author interview, June 23, 1995.

27. Pliny Fisk III, author interview, February 29, 1996.

28. E. S. Lipinsky, "Fuels from Biomass: Integration with Food and Materials Systems," *Science* 199 (February 10, 1978): 644.

29. See Pliny Fisk III, "A Sustainable Farm Demonstration for the State of Texas."

30. Pliny Fisk III, author interview, June 13, 2000.

31. Arnold Pacey, *Meaning in Technology,* p. 17.

32. Pliny Fisk III, author interview, February 29, 1996.

33. Ibid.

34. See Reuven Brenner, *History—The Human Gamble,* p. 134. I have previously investigated the hypothesis that serious incidents of anti-Semitic behavior took place at Blueprint Farm. See Moore, "Sustainable Technology and Landscapes of Exclusion: Methodology and the 'Left-out' Hypothesis." Based upon the field research conducted after the formation of the "left-out" anti-Semitic hypothesis, I

have concluded that, although some such incidents did indeed occur, they are not prominent, nor particularly salient to this investigation.

35. Brenner, *History—The Human Gamble,* p. 28. "Utopia, which was once a mere dream in the pre-industrial world, increasingly became a possibility with the development of modern technology. Today, I would insist that it has become a necessity—that is, if we are to survive the ravages of a totally irrational society that threatens to undermine the fundamentals of life on this planet."

36. "Celia Juárez," author interview, June 20, 1995.

37. "Thomas Rosas," author interview, June 23, 1995.

38. "Dr. Héctor Jiménez," author interview, June 21, 1995.

39. Ibid.

40. "Pat Roberts," author interview, October 10, 1995.

41. Ibid.

42. See Paul Thompson, "Sustainability Comes in Many Colors" (photocopy), p. viii; Gordon K. Douglass, *Agricultural Sustainability in a Changing World Order.*

43. David Harvey, *Justice, Nature & the Geography of Difference,* p. 182.

44. Paul Thompson, "Sustainability: What It Is and What It Is Not" (photocopy).

45. Campbell has produced the most coherent model of sustainability to date. He continues the triangulated model of competing interests used by Thompson and others. See Scott Campbell, "Green Cities, Growing Cities, Just Cities: Urban Planning and the Contradictions of Sustainable Development," *APA Journal,* Summer 1996, pp. 296–312.

46. See Simon Guy and Graham Farmer, "Re-interpreting Sustainable Architecture: The Place of Technology," *Journal of Architectural Education* 54, no. 3 (February 2001, forthcoming).

47. See Merritt Roe Smith, "Technological Determinism in American Culture."

48. For an example of technological voluntarism, see George Basalla, *The Evolution of Technology.*

49. "The simplest means of transforming the juxtaposed set of allies into a whole that acts as one is to tie the assembled forces *to one another,* that is, to build a machine. A machine, as its name implies, is first of all, a machination, a stratagem, and a kind of cunning; where borrowed forces keep one another in check so that nothing can fly apart from the group. This makes a machine a different thing from a tool which is a single element held *directly* in the hand of a man or a woman." Bruno Latour, *Science in Action,* p. 129.

50. "The word *black box* is used by cyberneticians whenever a piece of machinery or a set of commands is too complex. In its place they draw a little box about which they need to know nothing but its input and output. . . . That is, no matter how controversial their history, how complex their inner workings, how large the commercial or academic networks that hold them in place, only their input and output count." Ibid., p. 2.

The power of what Latour has conceived as the black box of technoscience is its ability to stop inquiry. Rather than prompting questions about how and why the box works, the overwhelming complexity of its contents prohibits examination for fear of discovering a bottomless pit. Rather than challenge such superiority, even the most skeptical will join the network because they see their own interests as complicit with the output of the box. In the end, Latour concludes that "When many elements are made to act as one, this is what I now call a black box." Ibid., p. 131. Many elements are made to work together as one when those who constructed it close the lid of the black box. When discourse about how and why things are done is blocked, all that remains is to minister to the needs of input and collect the benefits of output. The interest of the makers is invested in keeping the lid on. A *demonstration* of technology is about impressing witnesses with the operation of a black box. It is about winning converts, not about opening the lid to contemplate what is inside.

Chapter Five

1. In my view, Heidegger essentializes technology. He contends that the very essence of modern technology is that it has achieved a previously unknown autonomy from society. Where moderns have commonly adopted a *voluntarist* attitude toward technology—meaning that it is controlled by social processes— Heidegger adopts a *determinist* attitude—meaning that he understands modern technology to embody a trajectory that is independent of social processes. By Heidegger's account, modern technology leads to a diminished ontological condition. His concerns are less about the destruction of nature per se than about the human distress brought about by the technological understanding of Being. I want to argue that, although Heidegger's skepticism has philosophical merit, he failed to anticipate the degree to which contemporary technology might anticipate, and participate in, organic complexity.

2. See W. Bijker and J. Law, "General Introduction." The tradition of social constructivist theory in science and technology studies cannot be attributed to even a small number of authors. Bijker and Law are, however, among the most prominent.

3. Barry Barnes, "The Comparison of Belief Systems: Anomaly versus Falsehood."

4. See Alberto Pérez-Gómez, "Architecture as Science: Analogy or Disjunction." This observation derives from Pérez-Gómez's historical analysis of the relationship between architecture and science. See *Architecture and the Crisis of Modern Science.*

5. See Ruth Cowan, "How the Refrigerator Got Its Hum."

6. Ibid., p. 214.

7. For example, see Thomas P. Hughes, "Edison and Electric Light."

8. See Andrew Feenberg, *Critical Theory of Technology,* pp. 3, 12–17, 135–136, 175–176.

9. See Terry Winograd and Fernando Flores, *Understanding Computers and Cognition.*

10. For example, see Bruno Latour, *Science in Action,* pp. 53–54. Along with Michel Callon, Latour is generally considered to be the progenitor of actor-network theory.

11. Latour, *We Have Never Been Modern,* p. 95.

12. Latour, *Laboratory Life: The Construction of Scientific Facts,* p. 243.

13. See Latour, *We Have Never Been Modern,* pp. 15–35.

14. For example, see Merritt Roe Smith and Leo Marx, eds., *Does Technology Drive History?*

15. Andrew Feenberg, "Subversive Rationalization: Technology, Power, and Democracy," p. 7. See also Feenberg, *Critical Theory of Technology,* p. 121.

16. Donald MacKenzie and Judith Wajcman, eds., *The Social Shaping of Technology,* p. 6.

17. See Leo Marx and Merritt Roe Smith, "Introduction," in *Does Technology Drive History?*

18. Langdon Winner, "Do Artifacts Have Politics?," p. 26. See also *The Whale and the Reactor: a Search for Limits in an Age of High Technology* and *Autonomous Technology: Technics Out-of-Control as a Theme in Political Thought.*

19. See David Nye, *Consuming Power: A Social History of American Energies.*

20. Winner, "Do Artifacts Have Politics?," p. 37.

21. See Feenberg, "Subversive Rationalization: Technology, Power, and Democracy," p. 5.

22. Bijker and Law, "General Introduction," p. 3.

23. The same point is made by Andrew Pickering, "From Science as Knowledge to Science as Practice."

24. See T. J. Pinch and W. E. Bijker, "The Social Construction of Facts and Artifacts: Or How the Sociology of Science and the Science of Sociology of Technology Might Benefit Each Other," p. 29.

25. Ibid., p. 40.

26. Ibid., p. 44.

27. According to Donald MacKenzie and Judith Wajcman, *technology* cannot be understood as objects or hardware. Rather, they argue that technology has three dimensions: *human knowledge, patterns of human activity,* and *sets of objects.* See *The Social Shaping of Technology,* pp. 3–4.

28. In "The American Army and the M-16 Rifle," James Fallows documents a particularly tragic case where social form was considered more important than even the lives of soldiers.

29. Winner, "Do Artifacts Have Politics?," p. 37.

30. Before the Center for Maximum Potential Building Systems was commissioned by the Meadows Foundation to provide design and construction services for Blueprint Demonstration Farm, Cavazos & Associates of Laredo were retained to provide a schematic phase design proposal. This proposal was conventional in every sense. Private collection, S. Moore (1986?).

31. See Thomas Hughes, "Technological Momentum," p. 101.

32. The term "black box" is Latour's. See Latour, *Science in Action,* pp. 2, 81–82, 131.

33. Henri Lefebvre, *The Production of Space,* p. 113.

34. See Robert L. Thayer, Jr., *Gray World, Green Heart: Technology, Nature, and the Sustainable Landscape.*

35. Air-conditioning capacity is measured in the energy required to melt a ton of ice. In this sense, the weight of ice is commensurate with a volume of cool space.

36. This term belongs to Bruno Latour, *Science in Action,* p. 223.

37. This system never worked. Although problems were identified with the humidistat that controlled water supply to the Hexel pads, my own conclusion is that the design of the system was faulty. First, the design of this system relies upon a constant supply of low-humidity ambient air. In fact, the average humidity at Laredo is too high for the system to work reliably or consistently. Second, and more critically, the system designed for Blueprint Farm makes no attempt to utilize nocturnal long-wave radiative cooling of thermal mass. The traditional windtowers, or bod-girs, of medieval Iran relied primarily upon the storage of nocturnal coolness in masonry towers, not on evaporative cooling. For a full discussion of the antecedent technology, see: Mehdi Bahadoori, "Passive Cooling Systems in Iranian Architecture," *Scientific American,* 238, no. 2 (1978), pp. 144–151; "A Passive Cooling/Heating System for Hot-Arid Regions," p. 364; "Passive and Low-Energy Cooling," pp. 144–159.

38. See Sal Restivo, "Science, Sociology of Science, and the Anarchist Tradition," p. 36.

39. In "A Sustainable Farm Demonstration for the State of Texas," a paper delivered at the 1989 annual meeting of the American Solar Energy Society, Fisk argued that the technologies employed at Blueprint Farm derived from "local knowledge networks." There is, however, no suggestion that local farmworkers should, or did, play any role in conceptualizing how local knowledge might be transformed. This interpretation was confirmed in an interview with Pliny Fisk III, June 13, 2000.

40. Feenberg, *Critical Theory of Technology,* p. 118.

Chapter Six

1. For a very clear description of the sometimes inverted relationship between subjects and objects in contemporary life, see Charles Spinoza, Fernando Flores, and Hubert Dreyfus, *Disclosing New Worlds: Entrepreneurship, Democratic Action, and the Cultivation of Solidarity,* p. 9. The analysis of these authors contributes to Latour's definition of "quasi-objects."

2. See Robert Mugerauer, *Interpreting Environments: Tradition, Deconstruction, Hermeneutics,* pp. xxiv.

3. Robert Holub, *Reception Theory: A Critical Introduction,* p. 57.

4. For example, see Magali Sarfatti Larson, *Behind the Postmodern Facade: Architectural Changes in Late Twentieth Century America.*

5. Ibid., p. 71.

6. Henri Lefebvre, *The Production of Space,* p. 58.

7. Holub, *Reception Theory,* p. 2.

8. See Robert Holub, *Crossing Borders: Reception Theory, Poststructuralism, Deconstruction.* Quotation is from p. ix.

9. See Arnold Pacey, *Meaning in Technology,* pp. 69–75, for a helpful discussion on the historic relationship of science and alchemy. Pacey would likely agree that the ecologists in this case, and Fisk in particular, were closer to operating as alchemists than as scientists. The difference is that, unlike the Israelis and local

experts, Pacey would see the desire of the ecologists to participate in nature, rather than distance themselves from it, as a very productive attitude.

10. "Vera Sassoon," author interview, June 21, 1995.

11. "The trait in the character of technology that concerns us finally, then, is its stability. The lack of stability is frequently a target of the critics of technology. The charges extend from suggestions that the course of technology is precarious to assertions that technology is headed for self-destruction. A recurring criticism holds that technology suffers from a profound moral or spiritual defect. Technology dehumanizes and alienates us, it is said." Albert Borgmann, *Technology and the Character of Contemporary Life: A Philosophical Inquiry,* p. 144.

12. Rafael Longoria, author interview, February 2, 1996.

13. "Thomas Rosas," author interview, June 23, 1995.

14. According to Marx, "…all progress in capitalistic agriculture is a progress in the art, not only of robbing the labourer, but of robbing the soil; all progress in increasing the fertility of the soil for a given time, is a progress toward ruining the lasting sources of that fertility. The more a country starts its development on the foundations of modern industry, like the United States, for example, the more rapid is the process of destruction. Capitalist production, therefore, develops technology, and the developing together of various processes into a social whole, only by sapping the original sources of all wealth—the soil and the labourer." It is not likely that many will accuse Marx of being an ecologist. His definition of *nature* (or soil) was so similar to that of *capital* that it would fail to satisfy contemporary ecologist thought. His insight does, however, foreshadow the emergence of social ecology by equating social justice and ecological health. See Karl Marx, "Divisions of Labour and Manufacture," p. 416.

15. "Alvaro Lacayo," author interview, June 23, 1995.

16. Bernadini's critique of Blueprint Farm is likely derived from Kenneth Frampton, "Rappel à l'ordre."

17. "Rafael Bernadini," author interview, June 20, 1995.

18. "Pat Roberts," author interview, October 10, 1995.

19. According to Jim Hightower, "The basis of Land Grant teaching, research, and extension work has been that 'efficiency' is the greatest need in agriculture. Consequently, this agricultural complex has devoted the overwhelming share of its resources to mechanize all aspects of agricultural production and make it a capital intensive industry; to increase crop yield per acre through genetic manipulation and chemical application; and to encourage 'economies of scale' and vertical integration of the food process. It generally has aimed at transforming agriculture from a way of life to a business and a science, transferring effective control from the farmer to the business executive and the systems analyst." Jim Hightower, *Hard Tomatoes, Hard Times,* p. 2.

20. The political economist Reuven Brenner argues that all human activity can be related to the concept of "self-preservation." Under that umbrella, he has formulated two premises that frame history: the first premise is that inequality is what motivates people to be creative and take risks; the second premise is that

"markets, legal institutions and literacy emerge in response to increases in human population." The first premise explains the emergence of new ideas, and the second explains why ideas gain followers in one period rather than another. Brenner extends this logic to argue that when the material interests of people are challenged, and Jews are present, those marginalized by economic conditions will tend to "gamble" on anti-Semitic behavior. The same argument can be made with regard to homophobia and other forms of apparent bigotry. See Reuven Brenner, *History—The Human Gamble,* pp. ix, 127.

21. Richard Rorty has claimed that morality is contained within language. Those who don't "talk like us" become excluded from the protection of a shared ethical system. See Richard Rorty, *Contingency, Irony, and Solidarity,* p. 58.

22. "Pat Roberts," author interview, October 10, 1995.

23. See Lefebvre, *The Production of Space,* p. 117; Martin Heidegger, "The Age of the World Picture."

24. "Alvaro Lacayo," author interview, June 23, 1995.

25. Pacey, *Meaning in Technology,* p. 70.

26. Thomas McLaughlin, *Street Smarts and Critical Theory: Listening to the Vernacular,* p. 13.

Chapter Seven

1. Henri Lefebvre, *The Production of Space,* p. 59.

2. Jennifer Tann, "Space, Time, and Innovation Characteristics: The Contribution of Diffusion Process Theory to the History of Technology," *History of Technology* 17 (1995): 143–163.

3. Steven Shapin and Simon Schaeffer, *Leviathan and the Air-Pump: Hobbes, Boyle and the Experimental Life,* p. 77.

4. "Roberto Elizondo," author interview, June 24, 1995.

5. "Alvaro Lacayo," author interview, June 23, 1995.

6. "Dr. Héctor Jiménez," author interview, June 21, 1995.

7. Cited by Dr. Héctor Jiménez in ibid.

8. "Alvaro Lacayo," author interview, June 23, 1995.

9. Ibid.

10. "Pat Roberts," author interview, October 10, 1995.

11. See J. K. Gibson-Graham, *The End of Capitalism (As We Knew It): A Feminist Critique of Political Economy,* p. 260.

12. "Alvaro Lacayo," author interview, June 23, 1995.

13. Jim Hightower, author interview, November 21, 1995.

14. "Pat Roberts," author interview, October 10, 1995.

15. "Robert Deere," author interview, February 8, 1996.

16. The appropriation of a site presumes its redescription for new purposes. James Duncan has argued that every description and redescription of place has, in fact, two sites: the site *worked upon* and the site *represented.* The distinction lies in the intended, or unconsciously projected, transformation created by those who

do the describing. See James Duncan, "Sites of Representation: Place, Time, and the Discourse of the Other."

17. For example, see Robert Thayer, Jr., *Gray World, Green Heart: Technology, Nature, and the Sustainable Landscape,* 284; Patricia Leigh Brown, "Fisk Builds His Green House," *New York Times,* February 15, 1996. This *New York Times* article reviews Fisk's most recent project, the "Advanced Green Builder Demonstration Home," but makes reference to Blueprint Farm as his most ambitious project to date. The "Advanced Green Builder" project is also featured in Philip Jodidio, "The Center for Maximum Potential Building Systems: Advanced Green-Builder Demonstration, Austin, Texas," in *Contemporary American Architects,* vol. 4. In this volume, CMPBS's work is featured between that of Frank Gehry and Peter Eisenmann. See also Steve Lerner, "Pliny Fisk III: The Search for Low-Impact Building Materials and Techniques"; Pliny Fisk III and Richard MacMath, "Anybody There?: Architects, Design, and Responsibility," *Texas Architect,* November–December 1997, pp. 23–28; Pliny Fisk III, "Interview: Pliny the Greener," *Architecture,* June 1998, pp. 55–58; Pliny Fisk III, "Projects: Advanced Green Builder Demonstration Home, Laredo Blueprint Demonstration Farm, Noland Residence," *Perspecta* 29 (1998): 70–73.

18. Discussed by Bernard Beckerman in *Theatrical Presentation: Performer, Audience, and Act,* pp. x, xi.

19. David E. Nye, *American Technological Sublime,* p. 239.

20. Ibid., p. 222.

21. Lefebvre, *The Production of Space,* p. 133.

22. Donald MacKenzie and Judith Wajcman, *The Social Shaping of Technology,* p. 3.

23. Latour, *We Have Never Been Modern,* p. 117.

24. See David Brain, "Cultural Production as 'Sociology in the Making': Architecture as an Exemplar of the Social Construction of Cultural Artifacts," p. 197.

25. According to Richard Rorty, "anything can be made to look good or bad by being redescribed." The essential difference between "redescription" and "invention" is the ahistorical quality of inventing something. Before something can be redescribed, one must recognize the historical existence of the thing described. See Richard Rorty, *Contingency, Irony, and Solidarity,* p. 73.

Chapter Eight

1. See Clifford Geertz, *The Interpretation of Cultures: Selected Essays.*
2. Bruno Latour, *We Have Never Been Modern,* p. 117.
3. David Harvey, *Justice, Nature & the Geography of Difference,* p. 110.
4. Mike Davis, *City of Quartz.*
5. "Alvaro Lacayo," author interview, June 23, 1995.
6. "Celia Juárez," author interview, June 20, 1995.
7. "Dr. Héctor Jiménez," author interview, June 21, 1995.
8. Lefebvre, *The Production of Space,* p. 220.
9. Harvey, *Justice, Nature & the Geography of Difference,* p. 197.

10. Fredric Jameson, *Seeds of Time,* pp. 189–194.

11. Harris Sobin has demonstrated that Le Corbusier's attitude toward technology evolved through three stages: the "purist," or "high tech," period of the 1920's; the "transitional," or "reassessment," phase of the 1930's; and a "primitivist," or "low-tech," phase from 1945 to 1965. Plan Voisin, used as an exemplar here, emerged in the purist phase of Le Corbusier's relationship to technology. Sobin makes the significant claim that Le Corbusier's late work of the primitivist phase emerged out of his own recognition of the unintended consequences of the high-tech. Sobin further argues that Le Corbusier's primitivist experiments with ventilation, the "l'aerateur" in particular, are the precursors to contemporary concerns for environmentally responsible architecture. To argue that Le Corbusier's position is one inalterably opposed to local conditions is then something of a gross generalization. To argue, however, that Plan Voisin is an exemplar of purist technology is quite appropriate. See Harris Sobin, "From l'Air Exact to l'Aerateur: Ventilation and Its Evolution in the Architectural Work of Le Corbusier."

12. Philip Brey makes the far more sophisticated argument that "over the past two centuries, the role of geographic features in the constitution of the identity of places has decreased; this devaluation has resulted from the employment of various space-shaping technologies, used by human beings to transcend the limitations of their local environment." Yet even Brey's "geographical disembedding thesis" rests upon the ontological loss of traditionally geography-bound place. See Philip Brey, "Space-Shaping Technologies and the Disembedding of Place," p. 239.

13. Rem Koolhaas, *Delirious New York,* p. 212.

14. See Kurt Anderson, "Is Seaside Too Good to Be True?" Quotation from p. 46. See also Steven Moore, "The Disappearing Suburb," *Design Book Review* 26 (Fall 1992): 9–12.

15. Anderson, "Is Seaside Too Good to Be True?," p. 43.

16. Susan Stewart, with reference to Jacques Derrida, has defined "nostalgia" as "the repetition that mourns the inauthenticity of all repetition and denies the repetition's capacity to form identity." The longing of neotraditionalists and new urbanists for unique and authentic identity is realized in their well-intended, yet misguided, attempt to recoup the presumed grace of lost social structures. See Susan Stewart, *On Longing,* p. 24.

17. Although some will assume that this royal reference is to Prince Charles, see instead, Joel Barna, *The See-Through Years,* p. 21. Robert Davis, the tasteful developer of Seaside, embodies the developer-prince that Barna describes so well in his history of Texas development in the 1980's.

18. David Mohney and Keller Easterling, eds., *Seaside: Making a Town in America,* p. 71.

19. See Gianni Vattimo, *The End of Modernity: Nihilism and Hermeneutics in Postmodern Culture,* pp. 7–8, 100–103, 179.

20. Ibid., p. 166.

21. See Gevork Hartoonian, *Ontology of Construction,* pp. 11, 19.

22. See Rem Koolhaas, cited by Clifford A. Pearson, "Asian Cities: Is 'Generic' the Wave of the Future?," *Architectural Record,* March 1996, p. 19.

23. Koolhaas, *Delirious New York,* p. 51.

24. Ibid., p. 207.

25. For a full definition of the position now described as "eco-tech" see Catherine Slessor, *Eco-tech: Sustainable Architecture and High Technology.*

26. See Kenneth Frampton, *Modern Architecture: A Critical History,* p. 327.

Appendix: The Things Themselves

1. See Sal Restivo, "Science, Sociology of Science, and the Anarchist Tradition," p. 36.

2. See Kathryn Henderson, "The Battle over Building Codes: Straw Bale Building Moves into the Mainstream," presented at the Society for the Social Study of Science (4S) meetings October 1999, San Diego, California; "Do Building Codes Fertilize or Fumigate Grass Roots Technical Knowledge: How the Straw Bale Renaissance Negotiates Building Codes," presented at the Ninety-fourth Annual Meeting of the American Sociological Association, August 1999, Chicago, Illinois.

3. See Athena Swentzell Steen, *The Straw Bale House.*

4. Howard Reichmuth, PE, personal correspondence from Bend River, Oregon, to [Dr. Steven A. Moore, Austin, Texas] ALS, January 18, 2000, author's private collection, School of Architecture, University of Texas, Austin.

5. Pliny Fisk III, author interview, February 29, 1996.

6. See undated report by Howard Reichmuth, *An Assessment of Solar Zeolite Cooling in Laredo, Texas.*

7. See Ruth Cowan, "How the Refrigerator Got Its Hum."

8. Reichmuth to [Dr. Steven A. Moore, Austin, Texas] ALS, January, 18, 2000.

9. Tahal Consulting Engineers, Ltd., *Demonstration Farm; Laredo Junior College; Laredo, Texas, Evaluation and Planning Study,* p. D-7.

10. Ibid., p. D-4.

11. Ibid., p. D-5.

12. Ibid.

13. Ibid.

14. Pliny Fisk III, author interview, February 29, 1996.

15. See Steve Lerner, "John Todd: Greenhouse Treatment of Municipal Sewage."

REFERENCES

Adorno, Theodor W. "Fetish Character in Music and Regression of Listening." In *The Essential Frankfurt School Reader,* ed. Andrew Arato and Eike Gebhardt. New York: Continuum, 1994.

Agnew, John. *Place and Politics: The Geographical Mediation of State and Society.* Boston: Unwin & Allen, 1987.

——. "Representing Space: Space, Scale, and Culture in Social Science." In *Place/Culture/Representation,* ed. James Duncan and David Ley. New York: Routledge, 1993.

Alexander, Christopher. *A Pattern Language: Towns, Buildings, Constructions.* New York: Oxford University Press, 1977.

Althusser, Louis. *Politics and History: Montesquieu, Rousseau, Hegel and Marx.* Translated from the French by Ben Brewster. London: New Left Books, 1972.

Anderson, Kurt. "Is Seaside Too Good to Be True?" In *Seaside: Making a Town in America,* ed. David Mohney and Keller Easterling, pp. 42–47. New York: Princeton Architectural Press, 1991.

Anzaldúa, Gloria. *Borderlands La Frontera: The New Mestiza.* San Francisco: Spinsters/Aunt Lute, 1987.

Bahadoori, Mehdi. "Passive and Low Energy Cooling." In *Proceedings of Solar Prospects in the Arab World, Second Arab International Solar Energy Conference, Bahrain, 12–15 February,* ed. H. Alawi, A. Al-Jassar, F. Al-Juwayhel, A. Makafie, A. Al-Hamoud, S. Ayyash, and M. Kellow, pp. 144–159. Oxford: Pergamon Press, 1986.

——. "A Passive Cooling/Heating System for Hot-Arid Regions." In *Proceedings of Solar '88, the 13th Annual Passive Solar Conference of the American Solar Energy Society,* pp. 364–367. Boulder, Colo.: American Solar Energy Society, 1988.

———."Passive Cooling Systems in Iranian Architecture." *Scientific American* 238, no. 2 (1978): 144–151.

Barna, Joel. *The See-Through Years*. Houston: Rice University Press, 1992.

Barnes, Barry."The Comparison of Belief Systems: Anomaly versus Falsehood." 1973. Author's private collection, School of Architecture, University of Texas, Austin. Photocopy.

Basalla, George. *The Evolution of Technology*. Cambridge: Cambridge University Press, 1988.

Bayless, Lynn. "Strawbale and Steel." *Earthword: The Journal of Environmental and Social Responsibility* 1, no. 5 (1994): 36–38.

Beckerman, Bernard. *Theatrical Presentation: Performer, Audience, and Act*. New York: Routledge, 1990.

Berry, Wendell. *The Unsettling of America: Culture and Agriculture*. San Francisco: Sierra Club Books, 1977.

Bijker, W., and J. Law. "General Introduction." In *Shaping Technology/Building Society,* ed. W. Bijker and J. Law, pp. 1–20. Cambridge, Mass.: MIT Press, 1992.

Bimber, Bruce. "Three Faces of Technological Determinism." In *Does Technology Drive History?,* ed. Merritt Roe Smith and Leo Marx, pp. 79–100. Cambridge, Mass.: MIT Press, 1995.

Bookchin, Murray. *The Ecology of Freedom*. Montreal: Black Rose Books, 1991.

———. *The Limits of the City*. Montreal: Black Rose Books, 1973.

———. *The Modern Crisis*. Montreal: Black Rose Books, 1986.

———. *The Philosophy of Social Ecology*. Montreal: Black Rose Books, 1995.

———. *Post-Scarcity Anarchism*. Montreal: Black Rose Books, 1971.

———. *Toward an Ecological Society*. Montreal: Black Rose Books, 1980.

Borgmann, Albert. *Crossing the Postmodern Divide*. Chicago: University of Chicago Press, 1992.

———. *Technology and the Character of Contemporary Life: A Philosophical Inquiry*. Chicago: University of Chicago Press, 1984.

Brain, David. "Cultural Production as 'Sociology in the Making': Architecture as an Exemplar of the Social Construction of Cultural Artifacts." In *The Sociology of Culture: Emerging Theoretical Perspectives,* ed. Diane Crane, pp. 191–220. Cambridge, Mass.: Blackwell, 1994.

Bramwell, Anna. *Blood and Soil: Richard Walter Darre and Hitler's Green Party*. Abbottsbrook, England: Kensal House, 1985.

———. *Ecology in the Twentieth Century: A History*. New Haven, Conn.: Yale University Press, 1989.

Brenner, Reuven. *History—The Human Gamble*. Chicago: University of Chicago Press, 1983.

Brey, Philip. "Space-Shaping Technologies and the Disembedding of Place." In *Philosophy and Geography III: Philosophies of Place,* ed. Andrew Light and Jonathan Smith, pp. 239–263. New York: Rowman and Littlefield, 1998.

Brown, Patricia Leigh. "Fisk Builds His Green House." *New York Times,* February 15, 1996.

Butterfield, Box. "Nationwide Drop in Murders Is Reaching to Small Towns." *New York Times,* May 9, 2000. Available at http://archives.nytimes.com/archives/.

Campbell, Scott. "Green Cities, Growing Cities, Just Cities: Urban Planning and the Contradictions of Sustainable Development." *APA Journal,* Summer 1996, pp. 296–312.

Casey, Edward. *Getting Back into Place: Toward a Renewed Understanding of the Place-world.* Bloomington: University of Indiana Press, 1993.

Cochrane, Willard. *The Development of American Agriculture: A Historical Analysis.* Minneapolis: University of Minnesota Press, 1978.

Collins, Peter. *Changing Ideals in Modern Architecture, 1750–1950.* Montreal: McGill University Press, 1965.

Cowan, Ruth. "How the Refrigerator Got Its Hum." In *The Social Shaping of Technology,* ed. Donald MacKenzie and Judith Wajcman, pp. 202–218. Philadelphia: Open University Press, 1985.

Cunningham, William A., and T. Lewis Thompson. "Passive Cooling with Natural Draft Cooling Towers in Combination with Solar Chimneys." 1986(?). University of Arizona, Environmental Research Laboratory, 2601 East Airport Drive, Tucson, AZ 85706. Author's private collection, School of Architecture, University of Texas, Austin. Photocopy.

Dacamara, Kathleen. *Laredo on the Rio Grande.* San Antonio: Naylor, 1949.

Damisch, Hubert. *The Origin of Perspective.* Translated by John Goodman. Cambridge, Mass.: MIT Press, 1995.

Davis, Mike. *City of Quartz.* New York: Vintage, 1992.

de Buys, William Eno. *Enchantment and Exploitation: The Life and Hard Times of a New Mexico Mountain Range.* Albuquerque: University of New Mexico Press, 1985.

Deleuze, Gilles, and Felix Guattari. *A Thousand Plateaus: Capitalism and Schizophrenia.* London: Athlone Press, 1988.

Dillman, C. D. "Border Urbanization." In *Borderlands Sourcebook,* ed. E. Stoddard, R. Nostrand, and J. West. Norman: University of Oklahoma Press, 1993.

Dinnerstein, Leonard. *Anti-Semitism in America.* New York: Oxford University Press, 1994.

Douglass, Gordon K. *Agricultural Sustainability in a Changing World Order.* Boulder, Colo.: Westview Press, 1984.

Dreyfus, Hubert L. *Being-in-the-World: A Commentary on Heidegger's Being and Time, Division 1.* Cambridge, Mass.: MIT Press, 1991.

——. *Husserl, Intentionality and Cognitive Science.* Cambridge, Mass.: MIT Press, 1982.

Duncan, James. "Sites of Representation: Place, Time, and the Discourse of the Other." In *Place/Culture/Representation,* ed. James Duncan and David Ley, pp. 39–56. London: Routledge, 1993.

Edgerton, Samuel. *The Renaissance Discovery of Linear Perspective.* New York: Basic Books, 1975.

Elzinga, Aant. "Science as a Continuation of Politics by Other Means." In *Controversial Science,* ed. T. Brant, S. Fuller, and W. Lynch, pp. 127–153. Albany: State University of New York Press, 1993.

Entrikin, Nicholas. *The Betweenness of Place: Toward a Geography of Modernity.* Baltimore: Johns Hopkins University Press, 1991.

Fallows, James. "The American Army and the M-16 Rifle." In *The Social Shaping of Technology,* ed. Donald MacKenzie and Judith Wajcman, pp. 239–251. Philadelphia: Open University Press, 1985.

Farenthold, Sissy. "Commerce without Conscience." *CITE: The Architecture and Design Review of Houston* 30 (Spring–Summer 1993): 12–14.

Feenberg, Andrew. *Critical Theory of Technology.* New York: Oxford University Press, 1991.

———. "Subversive Rationalization: Technology, Power, and Democracy." In *Technology and the Politics of Knowledge,* ed. Andrew Feenberg and Alastair Hannay, pp. 3–22. Bloomington: Indiana University Press, 1995.

Feldstein, Dan. "House Approval of NAFTA Highway Predicted." *Houston Chronicle,* August 29, 1995.

Findlay, Dennis, ed. *Texas County Statistics.* Austin: Texas Department of Agriculture, 1985.

Fisk, Pliny, III. "Interview: Pliny the Greener." *Architecture,* June 1998, pp. 55–58.

———. "Projects: Advanced Green Builder Demonstration Home, Laredo Blueprint Demonstration Farm, Noland Residence." *Perspecta* 29 (1998): 70–73.

———. "A Sustainable Farm Demonstration for the State of Texas." In *Integration Compendium.* Austin: Center for Maximum Potential Building Systems, n.d. Author's private collection, School of Architecture, University of Texas, Austin. Photocopy.

———, and Richard MacMath. "Anybody There?: Architects, Design, and Responsibility." *Texas Architect,* November–December 1997, pp. 23–28.

Follesdal, Dagfinn. "Brentano and Husserl on Intentional Objects and Perception." In *Husserl, Intentionality and Cognitive Science,* ed. Hubert L. Dreyfus. Cambridge, Mass.: MIT Press, 1982.

Frampton, Kenneth. "Architecture and Critical Regionalism." *R.I.B.A./Trans.* 1, no. 3 (1983): 15–25.

———. "Critical Regionalism: Modern Architecture and Cultural Identity." In *Modern Architecture: A Critical History,* 2d ed., pp. 313–327. New York: Thames & Hudson, 1985.

———. "Critical Regionalism Revisited." In *Critical Regionalism: The Pomona Proceedings,* ed. Spyros Amourgis. Pomona, Calif.: College of Environmental Design, California State Polytechnic University, 1991.

———. *Modern Architecture: A Critical History,* 2d ed. New York: Thames & Hudson, 1985.

———. "Place-form and Cultural Identity." In *Design after Modernism: Beyond the Object,* ed. John Thackara, pp. 51–66. New York: Thames & Hudson, 1988.

———. "Prospects for a Critical Regionalism." *Perspecta: The Yale Architectural Journal* 20 (1983): 147–162.

———. "Rappel à l'ordre." In *Constancy and Change in Architecture,* ed. Malcolm Quantrill, pp. 3–22. College Station: Texas A&M University Press, 1991.

———. "Reflections on the Autonomy of Architecture: A Critique of Contemporary

Production." In *Out of Site,* ed. Diane Ghirardo, pp. 17–26. Seattle: Bay Press, 1991.

———. "Seven Points for the Millennium: An Untimely Manifesto." *Architectural Review* 206, no. 1233 (1999): 76–80.

———. *Studies in Tectonic Culture.* Cambridge, Mass.: MIT Press, 1995.

———. "Themes and Variations." Lecture delivered at the University of California, College of Environmental Design, January 27, 2000. Author's private collection, School of Architecture, University of Texas, Austin. Photocopy.

———. "Toward a Critical Regionalism: Six Points for an Architecture of Resistance." In *The Anti-Aesthetic: Essays on Postmodern Culture,* ed. Hal Foster, pp. 16–30. Seattle: Bay Press, 1983.

Gallison, Peter, and Emily Thompson, eds. *The Architecture of Science.* Cambridge, Mass.: MIT Press, 1999.

Garreau, Joel. *The Nine Nations of North America.* Boston: Houghton Mifflin, 1981.

Geertz, Clifford. *The Interpretation of Cultures: Selected Essays.* New York: Basic Books, 1973.

Gibson-Graham, J. K. *The End of Capitalism (As We Knew It): A Feminist Critique of Political Economy.* Malden, Mass.: Blackwell, 1996.

Guba, Egon, ed. *The Paradigm Dialogue.* Newbury Park, Calif.: Sage, 1990.

Gurwitsche, Aron. "Husserl's Theory of Intentionality and Consciousness." In *Husserl, Intentionality and Cognitive Science,* ed. Hubert L. Dreyfus, pp. 59–72. Cambridge, Mass.: MIT Press, 1982.

Guy, Simon, and Graham Farmer. "Re-interpreting Sustainable Architecture: The Place of Technology." *Journal of Architectural Education* 54, no. 3 (February 2001, forthcoming).

Haggard, Ken, and Greg McMillan. "First California Approved Straw-bale Construction." *Earthword: The Journal of Environmental and Social Responsibility* 1, no. 5 (1994): 38–41.

Hamilton, David. "Tradition, Preferences, and Postures in Applied Qualitative Research." In *Handbook of Qualitative Research,* ed. Norman Denzin and Yvonna Lincoln, pp. 60–69. Thousand Oaks, Calif.: Sage Publications, 1994.

Hartoonian, Gevork. *Ontology of Construction.* Cambridge: Cambridge University Press, 1994.

Harvey, David. *The Condition of Postmodernity.* Cambridge, Mass.: Blackwell, 1990.

———. *Justice, Nature & the Geography of Difference.* Cambridge, Mass.: Blackwell, 1996.

Heidegger, Martin. "The Age of the World Picture." In *The Question Concerning Technology and Other Essays,* pp. 115–154. Translated and with an introduction by William Lovitt. New York: Harper & Row, 1977.

———. *Being and Time.* Translated by John MacQuarrie and Edward Robinson. London: SCM Press, 1962.

———. "Building Dwelling Thinking." In *Poetry Language Thought,* pp. 145–161. Translated by Albert Hofstadter. New York: Harper & Row, 1971.

———. *The Question Concerning Technology and Other Essays.* Translated and with an introduction by William Lovitt. New York: Harper & Row, 1977.

Held, David. *Introduction to Critical Theory: Horkheimer to Habermas.* Berkeley: University of California Press, 1980.

Henderson, Kathryn. "The Battle over Building Codes: Straw Bale Building Moves into the Mainstream." Presented at the Society for the Social Study of Science (4S) meetings, October 1999, San Diego, California.

——. "Do Building Codes Fertilize or Fumigate Grass Roots Technical Knowledge: How the Straw Bale Renaissance Negotiates Building Codes." Presented at the Ninety-fourth Annual Meeting of the American Sociological Association, August 1999, Chicago, Illinois.

Herf, Jeffery. *Reactionary Modernism: Technology, Culture and Politics in Weimar and the Third Reich.* Cambridge: Cambridge University Press, 1984.

Hickman, Larry. *John Dewey's Pragmatic Technology.* Bloomington: University of Indiana Press, 1990.

Hightower, Jim. *Hard Tomatoes, Hard Times: The Original Hightower Report, Unexpurgated.* Cambridge, Mass.: Shenkman, 1978.

Holub, Robert. *Crossing Borders: Reception Theory, Poststructuralism, Deconstruction.* Madison: University of Wisconsin Press, 1992.

——. *Reception Theory: A Critical Introduction.* London: Methuen, 1984.

Horgan, Paul. *Great River: The Rio Grande in North American History.* Hanover, N.H.: Wesleyan University Press, 1984.

Hughes, Thomas. "Edison and Electric Light." In *The Social Shaping of Technology,* ed. Donald MacKenzie and Judith Wajcman, pp. 39–52. Philadelphia: Open University Press, 1985.

——. "Technological Momentum." In *Does Technology Drive History?,* ed. Merritt Roe Smith and Leo Marx, pp. 101–114. Cambridge, Mass.: MIT Press, 1995.

Ingersoll, Richard. "Second Nature: On the Bond of Ecology and Architecture." In *Reconstructing Architecture,* ed. Thomas Dutton and Lian Hurst Mann, pp. 119–157. Minneapolis: University of Minnesota Press, 1996.

Jameson, Fredric. *The Seeds of Time.* New York: Columbia University Press, 1994.

Jodidio, Philip. "The Center for Maximum Potential Building Systems: Advanced Green-Builder Demonstration, Austin, Texas." In *Contemporary American Architects,* vol. 4, pp. 46–53. New York: Taschen, 1998.

Johnston, R. J. *A Question of Place.* Cambridge, Mass.: Blackwell, 1991.

Koolhaas, Rem. *Delirious New York.* New York: Oxford University Press, 1978.

——, and Bruce Mau. *S, M, L, XL.* New York: Monacelli Press, 1995.

Larson, Magali Sarfatti. *Behind the Postmodern Facade: Architectural Changes in Late Twentieth Century America.* Berkeley: University of California Press, 1993.

Latour, Bruno. *Laboratory Life: The Construction of Scientific Facts.* Princeton, N.J.: Princeton University Press, 1986.

——. *Science in Action.* Cambridge, Mass.: Harvard University Press, 1987.

——. "Visualization and Cognition: Thinking with Eyes and Hands." In *Knowledge and Society: Studies in the Sociology of Culture Past and Present,* vol. 3, pp. 1–40. Greenwich, Conn.: JAI Press, 1986.

——. *We Have Never Been Modern.* Cambridge, Mass.: Harvard University Press, 1993.

Leach, Neil. *Rethinking Architecture: A Reader in Cultural Theory*. New York: Routledge, 1997.

Le Corbusier. "Guiding Principles of Town Planning." In *Programs and Manifestoes of 20th Century Architecture*, ed. Ulrich Conrads, pp. 89–94. Cambridge, Mass.: MIT Press, 1990.

Lefebvre, Henri. *The Critique of Everyday Life*, vol. 1. Translated by John Moore with a preface by Michel Trebitsch. New York: Verso, 1991.

——. *The Production of Space*. Translated by Donald Nicholson-Smith. Cambridge, Mass.: Blackwell Press, 1991.

Lerner, Steve. "John Todd: Greenhouse Treatment of Municipal Sewage." In *Eco-Pioneers: Practical Visionaries Solving Today's Environmental Problems*, pp. 47–66. Cambridge, Mass.: MIT Press, 1998.

——. "Pliny Fisk III: The Search for Low-Impact Building Materials and Techniques." In *Eco-Pioneers: Practical Visionaries Solving Today's Environmental Problems*, pp. 19–37. Cambridge, Mass.: MIT Press, 1998.

Lincoln, Yvonna, and Egon Guba. *Naturalistic Inquiry*. Newbury Park, Calif.: Sage, 1985.

Lindheim, Milton. *The Republic of the Rio Grande*. Waco, Tex.: W. M. Morrison, 1964.

Lipinsky, E. S. "Fuels from Biomass: Integration with Food and Materials Systems." *Science* 199 (February 10, 1978): 644–651.

Longoria, Rafael. "Disparity and Proximity." *CITE: The Architecture and Design Review of Houston* 30 (Spring–Summer 1993): 8–11.

Lyle, John Tillman. *Regenerative Design for Sustainable Development*. New York: Wiley, 1994.

Lynch, Michael. "The Externalized Retina: Selection and Mathematization in the Visual Documentation of Objects in the Life Sciences." *Human Sciences* 11 (1988): 201–234.

McCracken, Grant. *The Long Interview*. Qualitative Research Methods Series 13. Newbury Park, Calif.: Sage, 1988.

McDonough, William. *The Hannover Principles*. William McDonough Architects, 116 East 27th Street, New York, N.Y. 10017. 1992. Author's private collection, School of Architecture, University of Texas, Austin. Photocopy.

MacKenzie, Donald, and Judith Wajcman, eds. *The Social Shaping of Technology*. Philadelphia: Open University Press, 1985.

McLaughlin, Thomas. *Street Smarts and Critical Theory: Listening to the Vernacular*. Madison: University of Wisconsin Press, 1997.

Marx, Karl. "Divisions of Labour and Manufacture." In *Capital*. Vol. 1 of *The Marx-Engels Reader*, ed. Robert C. Tucker. New York: Norton, 1978.

Marx, Leo, and Merritt Roe Smith. "Introduction." In *Does Technology Drive History?*, ed. Merritt Roe Smith and Leo Marx, pp. ix–xv. Cambridge, Mass.: MIT Press, 1995.

Massumi, Brian. "Which Came First? The Individual or Society? Which Is the Chicken and Which Is the Egg?: The Political Economy of Belonging and the Logic of Relation." In *Anybody*, ed. Cynthia C. Davidson, pp. 175–188. Cambridge, Mass.: MIT Press and Anyone Corp., 1997.

Misa, Thomas. "Retrieving Sociotechnical Change from Technological Determinism." In *Does Technology Drive History?,* ed. Merritt Roe Smith and Leo Marx, pp. 115–142. Cambridge, Mass.: MIT Press, 1995.

Monhey, David, and Keller Easterling, eds. *Seaside: Making a Town in America.* New York: Princeton Architectural Press, 1991.

Moore, Steven A. "Book Review Essay." *Journal of Architectural Education,* Spring 2000, pp. 245–249.

——. "The Disappearing Suburb." *Design Book Review* 26 (Fall 1992): 9–12.

——. "Place and Technology as Related Developments in Architecture: The Case of Laredo, Texas." In *The Proceedings of the ACSA Bi-regional Annual Conference, Southeast and Southwest Regions, November 9–11.* Panel G: Concept of Critical Regionalism in "Place" Making. San Antonio: University of Texas, 1995.

——. "Reaction or Synthesis: The Modern Opposition of Space and Place in Architecture." In *Proceedings of the ACSA Annual Meeting at Boston, March 8–12,* pp. 533–538. Washington, D.C.: Associated Collegiate Schools of Architecture, 1996.

——. "Sustainable Technology and Landscapes of Exclusion: Methodology and the 'Left-out' Hypothesis." In *Proceedings of Testing Ground: Doctoral Symposium in Architectural History, Theory and Criticism, Harvard University Graduate School of Design, February 17 & 18,* pp. 1–19. Cambridge, Mass.: Harvard University Graduate School of Design, 1995.

——. "Technology, Place and the Nonmodern Thesis." *Journal of Architectural Education* 54, no. 3 (February 2001, forthcoming).

——. "Value and Regenerative Economy in Architecture." In *Proceedings of the ACSA Annual Meeting at Dallas, March 15–18, 1997,* pp. 544–551. Washington, D.C.: Associated Collegiate Schools of Architecture, 1997.

Morris, William, ed. *The American Heritage Dictionary of the English Language.* Boston: Houghton Mifflin Co., 1981.

Mugerauer, Robert. *Interpreting Environments: Tradition, Deconstruction, Hermeneutics.* Austin: University of Texas Press, 1995.

Mumford, Louis. *The South in Architecture.* New York: Harcourt Brace and Co., 1941.

Murillo, Hermalinda Aguirre. "History of Webb County." Master's thesis, Southwest State Teachers College, 1941. Collection of Laredo Public Library.

Naess, Arne. *Ecology, Community, and Lifestyle.* New York: Cambridge University Press, 1989.

Nye, David. *American Technological Sublime.* Cambridge, Mass.: MIT Press, 1994.

——. *Consuming Power: A Social History of American Energies.* Cambridge, Mass.: MIT Press, 1998.

Pacey, Arnold. *Meaning in Technology.* Cambridge, Mass.: MIT Press, 1999.

Pearson, Clifford. "Asian Cities: Is 'Generic' the Wave of the Future?" *Architectural Record,* March 1996, pp. 19–20.

Perdue, Peter C. "Technological Determinism in Agrarian Societies." In *Does Technology Drive History?,* ed. Merritt Roe Smith and Leo Marx, pp. 169–200. Cambridge, Mass.: MIT Press, 1995.

Pérez-Gómez, Alberto. *Architecture and the Crisis of Modern Science.* Cambridge, Mass.: MIT Press, 1983.

———. "Architecture as Science: Analogy or Disjunction?" In *The Architecture of Science,* ed. Peter Gallison and Emily Thompson, pp. 337–352. Cambridge, Mass.: MIT Press, 1999.

Pickering, Andrew. "From Science as Knowledge to Science as Practice." In *Science as Practice and Culture,* ed. Andrew Pickering, pp. 1–29. Chicago: University of Chicago Press, 1992.

Pinch, T. J., and W. E. Bijker. "The Social Construction of Facts and Artifacts: Or How the Sociology of Science and the Science of Sociology of Technology Might Benefit Each Other." In *The Social Construction of Technological Systems,* ed. W. Bijker, T. Hughes, and T. Pinch, pp. 17–50. Cambridge, Mass.: MIT Press, 1985.

Quantrill, Malcolm. *Constancy and Change in Architecture.* College Station: Texas A&M University Press, 1991.

Reichmuth, Howard, PE. *An Assessment of Solar Zeolite Cooling in Laredo, Texas.* Howard Reichmuth, 806 SW 5th Street, Corvallis, Oregon 97333. Undated (1988?). Author's private collection, School of Architecture, University of Texas, Austin. Photocopy.

———. Personal correspondence, from Bend River, Oregon, to [Dr. Steven A. Moore, Austin, Texas] ALS, January, 18, 2000. Author's private collection, School of Architecture, University of Texas, Austin.

Restivo, Sal. "Science, Sociology of Science, and the Anarchist Tradition." In *Controversial Science,* ed. T. Brant, S. Fuller, and W. Lynch, pp. 21–41. Albany: State University of New York Press, 1993.

Ricoeur, Paul. *History and Truth.* Evanston: University of Illinois Press, 1965.

Rorty, Richard. *Contingency, Irony, and Solidarity.* London: Cambridge University Press, 1989.

Rush, Richard R. *The Building Systems Integration Handbook.* Boston: Butterworth-AIA, 1986.

Schwandt, Thomas. "Paths to Inquiry in Social Disciplines: Scientific, Constructivist, and Critical Theory Methodologies." In *The Paradigm Dialogue,* ed. Egon Guba, pp. 258–276. Newbury Park, Calif.: Sage, 1990.

Seamon, David. *A Geography of the Lifeworld: Movement, Rest, and Encounter.* New York: St. Martin's Press, 1979.

Shapin, Steven, and Simon Schaeffer. *Leviathan and the Air Pump: Hobbes, Boyle and the Experimental Life.* Princeton, N.J.: Princeton University Press, 1985.

Sheppard, Paul. "The Perils of the Green Revolution." In *Hard Tomatoes, Hard Times,* ed. Jim Hightower, pp. 315–325. Cambridge, Mass.: Shenkman, 1978.

Slessor, Catherine. *Eco-Tech: Sustainable Architecture and High Technology.* New York: McGraw-Hill, 1997.

Smith, Justin. "La Republica Rio Grande." *American History Review* 25, no. 4 (1920): 660–668.

Smith, Merritt Roe. "Technological Determinism in American Culture." In *Does Technology Drive History?,* ed. Merritt Roe Smith and Leo Marx, pp. 1–36. Cambridge, Mass.: MIT Press, 1995.

———, and Leo Marx, eds. *Does Technology Drive History?* Cambridge, Mass.: MIT Press, 1995.

Sobin, Harris. "From l'Air Exact to l'Aerateur: Ventilation and Its Evolution in the Architectural Work of Le Corbusier." In *Proceedings of the 84th ACSA Annual Meeting & Technology Conference at Boston, March 8–12, 1996,* pp. 220–227. Washington, D.C.: Associated Collegiate Schools of Architecture, 1996.

Soja, Edward. *Postmodern Geographies: The Reassertion of Space in Critical Social Theory.* London: Verso, 1989.

Spence, Karen Cordes. "Theorizing in Recent Architecture: An Examination of Texts by Frampton, Rossi and Lang." Doctoral dissertation, Texas A&M University, 1996. Collection of Evans Library, College Station, Texas.

Spinoza, Charles, Fernando Flores, and Hubert L. Dreyfus. *Disclosing New Worlds: Entrepreneurship, Democratic Action, and the Cultivation of Solidarity.* Cambridge, Mass.: MIT Press, 1997.

Steen, Athena Swentzell. *The Straw Bale House.* White River Junction, Vt.: Chelsea Green Publishers, 1994.

Stewart, Susan. *On Longing.* Durham, N.C.: Duke University Press, 1993.

Stilgoe, John. *Common Landscapes of America: 1580 to 1845.* New Haven, Conn.: Yale University Press, 1982.

Strauss, A., and J. Corbin. *Basics of Qualitative Research.* Newbury Park, Calif.: Sage, 1987.

Tafuri, Manfredo. *The Sphere and the Labyrinth: Avant-Gardes and Architecture from Piranesi to the 1970's.* Cambridge, Mass.: MIT Press, 1990.

Tahal Consulting Engineers, Ltd. *Demonstration Farm; Laredo Junior College; Laredo, Texas, Evaluation and Planning Study.* Tel Aviv, Israel: Jewish National Fund (Keren Kaymeth Leisrael) Inc., 1987. Photocopy.

Tann, Jennifer. "Space, Time, and Innovation Characteristics: The Contribution of Diffusion Process Theory to the History of Technology." *History of Technology* 17 (1995): 143–163.

Texas Department of Agriculture. "The Dissemination of Innovation: Training and Educational Development on the 'Blueprint' Demonstration Farm, Laredo, Texas. A Proposal Submitted to the Hitachi Foundation October, 1998." Texas Department of Agriculture File (TDAF) 4. Author's private collection, School of Architecture, University of Texas, Austin. Photocopy.

———. "Internal TDA Assessment of the Laredo Situation." Texas Department of Agriculture File (TDAF) 12.1, n.d. Author's private collection, School of Architecture, University of Texas, Austin. Photocopy.

———. "Mission Statement for the Laredo Blueprint Demonstration Farm," draft of March 1, 1988. Texas Department of Agriculture File (TDAF) 7.12. Author's private collection, School of Architecture, University of Texas, Austin. Photocopy.

———. "The Texas-Israel Exchange." Texas Department of Agriculture File (TDAF) 8.1. Author's private collection, School of Architecture, University of Texas, Austin. Photocopy.

Thayer, Robert, Jr. *Gray World, Green Heart: Technology, Nature, and the Sustainable Landscape.* New York: Wiley, 1994.

Thompson, Jerry. "Historical Survey." In *A Shared Experience: The History, Architec-*

ture and Historic Designations of the Lower Rio Grande Heritage Corridor, 2d ed., ed. Mario Sánchez, Ph.D., R.A., pp. 17–78. Austin: Los Caminos del Rio Heritage Project, Texas Historical Commission, 1994.

———. Sabers on the Rio Grande. Austin: Presidial Press, 1974.

Thompson, Paul. "Agrarianism and the American Philosophical Tradition." Agriculture and Human Values 7, no. 1 (1990): 3–8.

———. "Sustainability: What It Is and What It Is Not." 1998. Author's private collection, School of Architecture, University of Texas, Austin. Photocopy.

———. "Sustainability Comes in Many Colors." 1996. Author's private collection, School of Architecture, University of Texas, Austin. Photocopy.

Tilley, Ray Don. "Blueprint for Survival." Architecture, May 1991, pp. 64–69.

Tonnies, Ferdinand. Community and Society. New York: Harper, 1963.

Tzonis, Alexander, and Liane Lefaivre. "Critical Regionalism." In Critical Regionalism: The Pomona Proceedings, ed. Spyros Amourgis, pp. 3–28. Pomona, Calif.: College of Environmental Design, California State Polytechnic University, 1991.

———. "The Grid and the Pathway." Architecture in Greece, no. 5 (1981): 164–178.

Vattimo, Gianni. The End of Modernity: Nihilism and Hermeneutics in Postmodern Culture. Baltimore: Johns Hopkins University Press, 1985.

Walsh, Sister Natalie. "The Founding of Laredo and St. Augustine Church." Master's thesis, University of Texas at Austin, 1935. Collection of Laredo Public Library.

Wilkinson, J. B. Laredo and the Rio Grande Frontier. Austin: Jenkins Publishing Co., 1975.

Wilson, Patricia. Exports and Local Development: Mexico's New Maquiladora. Austin: University of Texas Press, 1992.

Winner, Langdon. Autonomous Technology: Technics Out-of-Control as a Theme in Political Thought. Cambridge, Mass.: MIT Press, 1977.

———. "Do Artifacts Have Politics?" In The Social Shaping of Technology, ed. Donald MacKenzie and Judith Wajcman, pp. 26–38. Philadelphia: Open University Press, 1985.

———. The Whale and the Reactor: A Search for Limits in an Age of High Technology. Chicago: University of Chicago Press, 1986.

Winograd, Terry, and Fernando Flores. Understanding Computers and Cognition. Reading, Mass.: Addison-Wesley, 1987.

INDEX

gateway to Mexico, 75. *See also* Laredo
(city of)
Gemeinschaft, 12
geography
definition of, 53
of technology, 75, 77, 81
Gesellschaft, 12
Gibson-Graham, J. K., 84
great man theory of history, 114, 126
greenhouse(s), 97, 124, 210, 214–215
Gurwitsche, Aaron, 88–89

H

habitation, 111, 138
Hartoonian, Gevork, 194
Harvey, David, 1–2, 6–8, 10–13, 19, 107,
182, 185–186
Heidegger, Martin, xiv, 1–2, 6–8, 10,
12–13, 19, 33, 46–47, 87, 89–92,
108–109, 111, 113, 116–120, 138,
146, 148, 184, 186, 194
Hightower, Jim, xviii, 26, 29–33, 40, 42–
43, 45, 59, 60, 62, 70, 72, 92–97, 99,
104–106, 108, 114, 123–125, 133,
144–145, 147, 160–164, 168, 170,
172, 176–180, 182, 209
Hightower network, 92, 93–96, 99, 105,
110, 123, 125, 133, 139, 176
history (definition of), 53
Holub, Robert, 135, 137–139, 148, 151.
See also reception theory
Hughes, Thomas, 116, 127, 130, 134.
See also systems theory
human agreements, 116, 119–120,
134, 180–182, 200
human liberation, 9–10, 15, 21, 174
human practices, 8, 24, 27, 50, 52, 74,
84, 87, 103, 111, 127, 177
Husserl, Edmund, 87–89, 109, 148. *See
also* intentionality

I

Icono-metrics, 165–167, 185, 204. *See
also* Center for Maximum Potential Building Systems
Ingersoll, Richard, 19–20. *See also* social ecology
inorganic mulch, 159, 213

instrumentalism, xii, xv, 35, 93, 95–96,
101, 116–117, 147, 149, 151, 154,
184, 189, 196
intentionality (theory of), 26, 87–90, 92,
108, 111–112, 148. *See also*
Heidegger, Martin; Husserl,
Edmund
interdisciplinary studies, xi
interpretive networks, 139–140. *See
also* reception theory
interpretive paradigm, 136, 148–149,
151, 153–154, 176
ecological, 154
formalist, 149–151, 154
market-driven, 148, 150–151
noncapitalist, 145, 150
interviews (data collection method), xviii
Israel(i), xii, 3–5, 10, 31–33, 36–40, 42–
45, 72–73, 92–98, 102–105, 111,
123–125, 139–152, 155, 159–164,
170–172, 176–177, 179–180, 203,
209, 212–216
network, 92, 110, 123, 172

J

Jackson, Jesse, 30–31, 42, 93, 168
Jameson, Fredric, xiii, 21–22, 187, 197–
198
Jauss, Hans Robert, 24, 136–139, 148
Jewish community, 31–33, 72, 93, 96–
97, 124, 168
Jewish network, 96–98, 104, 123, 133

K

Kahn, Louis, 34, 114, 118, 143
Katz-Oz, Avraham, 31
kibbutz, xii, 31, 33, 45, 124, 152, 160
Koolhaas, Rem, 10, 28, 189, 194–195,
197. *See also* nihilism

L

La Frontera Chica. *See* Frontera Chica
land grant complex, 60, 97
land grant network, 92, 94, 97, 104–
108, 110, 123, 125–127, 139, 142,
144–145, 148, 151, 162–166, 172,
176–177, 179, 182
Landschaft, 77–79

of), 2, 6–7, 10, 13–15, 18–19, 21–
    23, 28, 188. *See also* modern;
    nonmodern
practical-instrumental research, 95
practice (in relation to theory), xvii
pseudonyms, xviii
public show, 167–168

Q

quasi-object, 48, 90–91, 118, 187. *See
    also* Latour, Bruno
quasi-subject, 48, 90–91, 187. *See also*
    Latour, Bruno

R

racism, 12, 30
radical nihilism, 28, 174, 193. *See also*
    Koolhaas, Rem; Vattimo, Gianni
rainwater catchment system, 219
reception theory, 136–139. *See also*
    Holub, Robert; Jauss, Hans Robert
redescription, 164, 168–169, 172
regenerative architecture, xv–xvi, 8–9,
    18–19, 28, 130–134, 174, 177–182,
    184–188, 193, 197–201
    definition of, 8
regionalism
    devaluation of, 11–15
    romantic, xvi, 18, 69, 81, 143, 190.
        See also critical regionalism
Reichmuth, Howard, 206, 208–209, 219
reproduction, 8, 13, 28, 85, 116, 136,
    138, 154–159, 162–166, 168, 172–
    173, 184–185, 191, 200–201, 216
Rio Grande International Study Center
    (RISC), 44, 70, 165
Rorty, Richard, 13

S

science and technology studies (STS),
    xxii, 27, 112–117, 119, 126, 204
Seaside (Florida), 191–193. *See also*
    Duany, Andres; Platter-Zyberk,
    Elizabeth
self-determination, 134
self-realization, 9–10, 174–175, 186
self-sufficiency, 161
Semper, Gottfried, 25

sense of place, 47–49, 51, 65, 74, 81–
    84, 111, 179, 199. *See also* Agnew,
    John
settlement pond, 217–218, 220
shade structure(s), 178, 215–216
Smith, Merritt Roe, 109, 120
social constructivist theory, xiv, 47–48,
    114–116, 119–120, 122. *See also*
    Cowan, Ruth Schwartz
Social Darwinism, 62
social ecology, 18–21, 98, 107, 176,
    197. *See also* Bookchin, Murray
Soja, Edward, 12
solar
    dryer, 208–209
    hot water heater, 216
    zeolite refrigeration, 178–179, 205–
        208, 219
solid waste composting, 4, 40, 127,
    152, 170, 178, 210–211
space
    absolute, 81
    abstract, 10, 77, 81, 158
    binary, 78–80, 83
    liberative, 9, 65
    monistic, 78–80
    use and definition of, 46–49
Spanish Colonial Orders for Discovery
    and Settlement, 55–57, 74–78
Stilgoe, John, 77
straw-bale (construction), 127, 133,
    180, 185, 204–205
sublimity, 28, 103, 146, 156, 167–172,
    178–179, 200. *See also* Nye, David
sustainability, 27–28, 42–43, 96, 98,
    106, 123, 125, 139, 151–152, 162,
    164, 167–168, 176, 187, 197, 203
    definitions of, 20, 94–95, 98, 104,
    106–108, 156, 161–162, 164, 172,
    181
sustainable
    agriculture, 36
    development, 94
symbolic-instrumental research, 95
systems theory, 114, 116, 119, 126. *See
    also* Hughes, Thomas